ESSENTIALS OF
INFERENTIAL STATISTICS

Fifth Edition

Malcolm O. Asadoorian
Demetri Kantarelis

University Press of America,® Inc.
Lanham · Boulder · New York · Toronto · Plymouth, UK

Copyright © 2009 by
University Press of America,® Inc.
4501 Forbes Boulevard
Suite 200
Lanham, Maryland 20706
UPA Acquisitions Department (301) 459-3366

Estover Road
Plymouth PL6 7PY
United Kingdom

Library of Congress Control Number: 2008938694
ISBN-13: 978-0-7618-4451-8 (paperback : alk. paper)
ISBN-10: 0-7618-4451-1 (paperback : alk. paper)
eISBN-13: 978-0-7618-4452-5
eISBN-10: 0-7618-4452-X

To My Family
-Malcolm O. Asadoorian

To My Mother
-Demetri Kantarelis

"There are three kinds of lies: lies, damned lies, and statistics."
-Benjamin Disraeli

"It is easy to lie with statistics, but it is easier to lie without them."
-Frederick Mosteller

Contents

Preface   vii
Acknowledgements   ix
Introduction   xi

Chapter 1: Basic Definitions and Introduction to Probability

1.1 Basic Definitions   1
1.2 Introduction to Probability   5
1.3 Screening Tests, Epidemiological Tests and Bayesian Analysis   24

Chapter 2: Probability Distributions, Summary Measures, and Graphs

2.1 Probability Distributions   31
2.2 Population Parameters and Sample Statistics   34
2.3 Measures of Central Tendency and Variation   35
2.4 Random Variables   45
2.5 General Questions and Answers   58
2.6 Health and Medical Questions and Answers   66
2.7 Expected Value, Variance, Covariance
        and Correlation Questions and Answers   69

Chapter 3: Sampling Distributions and Interval Estimation

3.1 Sampling Distributions   81
3.2 Continuous Probability Distributions   89
3.3 Confidence Interval for the Mean of Population   100
3.4 Confidence Interval for the Population Proportion   108
3.5 Confidence Interval for the Variance of Population   108
3.6 Confidence Intervals for Two Population Parameters   110
3.7 Confidence Intervals for Health and Medical Applications   113
3.8 Sample Size   114
3.9 General Questions and Answers   119
3.10 Health and Medical Questions and Answers   127

Chapter 4: Hypothesis Testing

4.1 Hypothesis Testing Procedure   131
4.2 Two Types of Errors   138
4.3 Tests of Hypotheses for Two Population Parameters   142
4.4 Tests of Hypotheses for Health and Medical Applications   146
4.5 Sample Size for Specific Level of Significance and Power   148
4.6 Diagnostic/Lab Tests versus Hypothesis Tests   150

4.7 General Questions and Answers   151
4.8 Health and Medical Questions and Answers   156

Chapter 5: Correlation and Linear Regression

5.1 Correlation   161
5.2 Regression   165
5.3 Tests of Statistical Significance   172
5.4 General Questions and Answers   178
5.5 Health and Medical Questions and Answers   185

Chapter 6: Nonparametric Tests

6.1 The Lilliefors Test for Normality   190
6.2 The Wilcoxon Rank-Sum or Mann-Whitney Test   191
6.3 The Sign Test   194
6.4 The Wilcoxon Signed-Rank Test   198
6.5 The Kruskal-Wallis Test   202
6.6 The Spearman's Rank-Correlation Test   204
6.7 General Questions and Answers   207
6.8 Health and Medical Questions and Answers   213

Appendix I: Projects
    1. Automobile Mileage   218
    2. Pharmacological Methodology   229
    3. Chuck-A-Luck Game   231
    4. Sample Testing   233
    5. College Student Survey   239

Appendix II: Summation Notation   263

Appendix III: Tables
    1. Z Table   265
    2. T Table   267
    3. χ^2 Table   268
    4. F Table   270
    5. Durbin-Watson Table   274

Appendix IV: Quick-Reference Guides   275

References   281

Subject Index   285

Preface

Motivation for This Book

"*Essentials of Inferential Statistics*" emerged as an idea through years of teaching one semester general statistics courses. Although a brief introductory textbook that comprehensively covered the *key* essentials of inferential statistics would be ideally suited for audiences in these courses, such a text was not available in the market. Instead, students had to purchase the conventional two-semester "encyclopedic-type" textbook and accompanying study guide. Students not only had to carry around this bulky load, but also paid a high price for it. Thus, we hope that this brief book better satisfies the needs of the one semester general statistics student.

More specifically, our goal is to help all students who seek a general understanding of the essentials of inferential statistics. The book can also serve as a "skills refresher and updating tool" for those who completed a statistics course in the past. As such, it may be helpful to all professionals who desire to incorporate the key principles of statistical inference in their scientific endeavors.

Organization of Topics and Special Features

Statistical essentials and their applications, are demonstrated in 6 chapters. The first chapter stresses the basics of probability theory; the following chapters outline the basic mathematical tools and computing procedures for inferential decision-making. In order to minimize computational uncertainty, every chapter contains solved examples as well as questions with corresponding answers. Moreover, the use of any statistics software available in the market is recommended.

In addition to providing the key essentials of inferential reasoning, we provide a list of references at the end of the book to allow readers to pursue more detailed studies in accordance with their interests. With these references, this book can also serve as the core text for a seminar in inferential statistics. Likewise, the book may be used as a helpful supplement for any course that requires knowledge of the essentials of inferential decision-making.

New Features in the Fifth Edition

The major change in the fifth edition of this text is the addition of the basic properties of expectations and variance and their application to the field of managerial finance and investments. In addition, errors in terms of both form and content in the previous edition were corrected.

Acknowledgements

In preparing this book, invaluable help and comments were received from many people. We wish to thank Kevin Hickey, Tim Heffernan, and Neil Rankin (Assumption College); Mac Hill (Worcester State College); Randy Brill, George Reed, and Donald Tipper (University of Massachusetts, Medical School); and Larry Recht (Memorial Hospital, Worcester, MA). Grateful acknowledgement goes to the entire professional and administrative staff of the Neonatology Department at University of Massachusetts Memorial Healthcare, Memorial Campus of Worcester Massachusetts, for attending a Biostatistics seminar, and suffering through earlier drafts of this book. Above all, we would like to thank our statistics students at Assumption College and Lynn University, who have provided not only inspiration for writing this book, but also many quantitative and qualitative corrections and suggestions.

To our Chief Contributor

This text revision would have not been possible without the support and contribution of an *outstanding* student of statistics and economics named Shelley Hesselton-Mangan. Shelley sacrificed great time and effort to improve the quality of this text. To her, we offer a special thank you!

Introduction

As you read this book, you may have questions which only we can answer. You may also have suggestions and comments on how to improve this book. We would be delighted to correspond with you! Please feel free to contact us by e-mail, fax, telephone, or regular mail:

E-mail:	*masadoorian@lynn.edu*	*dkantar@assumption.edu*
Fax:	*561-237-7014*	*508-767-7382*
Tel:	*561-237-7755*	*508-767-7557*
Adrss:	*Dr. Malcolm O. Asadoorian III*	*Dr. Demetri Kantarelis*
	Lynn University	*Assumption College*
	College of Arts & Sciences;	*Department of Economics*
	College of Business & Management	*& Global Studies*
	3601 North Military Trail	*500 Salisbury Street*
	Boca Raton, FL 33431-5598	*Worcester, MA 01609*

Let us also take the opportunity to express our opinion on how you might best be able to learn statistics. Unfortunately, reading this or any textbook would not be enough! You need more than theory and conveniently designed examples. We strongly believe that you have to be involved in real world applications of statistics, so that you experience the difficulties associated with the "design" of statistics projects, "data collection", and "inferential interpretation". Thus, we wish to encourage you to design a statistics project, based on a real world application, and write a paper about it. We are certain your statistics teacher will be able to give you ideas. If that is not possible, please feel free to contact us! We will gladly brainstorm with you about possible topics.

It may be possible to publish an anthology of selected papers written by statistics students, and make it available as a supplement for this book. Please feel free to submit a typed version of your paper for such a possibility. If your paper gets accepted we will send you format instructions.

Malcolm O. Asadoorian, Ph.D.& Demetri Kantarelis, Ph.D.

Chapter 1

BASIC DEFINITIONS
AND
INTRODUCTION TO PROBABILITY

"Statistics is the science of inference"

1.1   Basic Definitions

1.1.1   Descriptive versus Inferential Statistics

In general, *statistics* is defined as the science of collecting, organizing, analyzing, and interpreting data in order to make decisions. Moreover, there are two main branches of statistics: *descriptive* and *inferential/inductive* statistics. *Descriptive Statistics* involves the organization, summarization, and display of data; *Inferential/Inductive Statistics* utilizes probabilistic techniques to analyze sample information from a certain population (known part), to improve our knowledge about the population (unknown whole). As indicated by the title of this text, the focus here is on inferential/inductive Statistics. However, *this text is unique in that we will be integrating descriptive statistics within the context of inferential statistics.* The following flow chart summarizes inferential statistics:

Flow Chart : Inferential Statistics

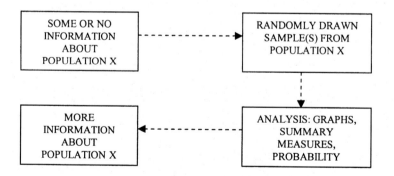

1.1.2   Understanding Data

Data is defined as information from observations, counts, measurements, or responses. Consider the following statements based on the collection of data:

"A . . . survey of traffic deaths during this past Memorial Day weekend shows a 36% decline in fatalities compared with last year" (*National Safety Council*)."

" . . . men who eat just two servings of raw tomatoes a week have a 34% lower risk of developing prostate cancer" (*Journal of the National Cancer Institute*)."

There exist two main types of data: a *population* and a *sample*. A population consists of *all* possible outcomes, responses, measurements, or counts (finite or infinite) in which the researcher is interested. It is a set of beings, creatures, objects, persons, things, or, in general, entities having some common measurable characteristic. This *set* is a variable which may be codified to represent sex, color, race, ethnicity, etc.

A sample is a *sub-set* of possibilities selected from the population. Sampling from the population is done randomly, such that every possibility of the population has the same likelihood of selection and different possibilities are selected independently. Samples selected this way are called *random* or *representative samples*. (sampling techniques will be discussed in more detail later).

In order to understand the difference between a population and sample, consider the following example. In a recent survey, 3000 American adults were asked if they read news on the Internet at least once a week, of which, 600 said "yes." The population in this example would be the responses of *all* American adults; the sample represents the 3000 American adults surveyed. Graphically, we can illustrate the population and sample by employing the mathematical construct known as the *Ballentine-Venn Diagram*:

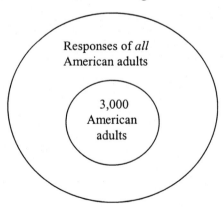

Responses of *all* American adults

3,000 American adults

The Diagram represents the population as the *set* of all responses and the sample as the *sub-set* of the population.

1.1.3   Statistic versus Parameter

A very important, and sometimes neglected, distinction is the difference between a *statistic* and a *parameter*. A statistic is a numerical/quantitative description of a sample; a parameter is a numerical/quantitative description of a population. Consider the following statements that illustrate the difference between the two:

Note: the symbol "$\Rightarrow$" means "implies"

"The average annual salary for 35 of a company's 1,200 accountants is $57,000" $\Rightarrow$ a statistic.

"In 1997, the interest category for 12% of all new magazines was sports" $\Rightarrow$ a parameter.

1.1.4   Experimental Design

The essential elements in designing a statistical research project are as follows:

Step A: Identify the variable(s) of interest (i.e. the focus) and the population of the study.

Step B: Develop a plan for collecting the data (discussed in detail in the next section).

Step C: Collect the data (the question of how much data will be discussed later in the text).

Step D: Employ the techniques of Inferential Statistics

1.1.5   Data Collection

The following options are typical in order to collect data:

Option A: Perform an *Experiment* (e.g. natural science, placebo and control group, etc.).

Option B: *Simulation* techniques in which computers are used to reproduce the conditions of a situation or process.

Option C: A *Census* which is defined as a count/measure of the entire population. It is important to note, however, that the U.S. Census of Population and Housing is actually based on a sub-set of the population (i.e. a sample).

Option D: *Sampling* techniques, which can be classified as five main types:

   i)   *Random* sampling, in which each member of a population has an equal chance of being selected. The random selection can be done typically via use of a computer. The following example represents a random sample. Assume we wish to sample households in a particular county. We can assign each household a number and then choose a sample by generating random numbers from computer software such as a Spreadsheet package. If the random numbers are: 2, 7, 9, and 13, we choose the households with those assigned numbers.

   ii) *Stratified* sampling divides the population into at least two (2) different sub-sets, commonly referred to as *strata* that share similar characteristics; a random sample is then selected from each. Using the same example of the households, we may divide them according to income: low income, middle income, and high income. However, in doing so, the researcher must predefine the ranges (i.e. the upper and lower limits) for each stratum.

   iii) *Clustering* divides the population into groups commonly referred to as *clusters* and then selecting all of the members in one or more, but not all, clusters. Again using our "running example" of the households, we may divide the population of all households in the United States according to geographic region: north, south, east, and west, and then choose one or more, but not all, regions.

   iv) *Systematic* sampling involves ordering the population in some way (e.g. income) and then selecting members of the population at regular/fixed intervals. For example, we may arrange the households in the United States from lowest income (on the left) to highest income (on the right) and then choose, say, every other household.

   v)   *Convenience* sampling involves using any members of a population that are readily available. This method is the LEAST preferred in that it is not random (i.e. there does not exist an equal chance of being chosen). An example of this technique is a survey that chooses to interview only people at a shopping mall. This is not a random sample because the likelihood of being selected is contingent upon the likelihood of going to the shopping mall!

1.2 Introduction to Probability

1.2.1   Probability Theory

Probability theory is the foundation of inferential statistics. 'Probability,' although widely used, is difficult to define. By 'probability' some imply proportion; others imply degree of certainty, others use it to amplify what is possible or impossible; moreover others think about it as odds in favor of (or odds against) an event. For example, consider the following set of 7 letters and questions and answers:

$$K, R, A, Z, R, H, K$$

Question 1: What is the probability or proportion of Ks?

Answer: 2/7

Question 2: Assume that the letters are drawn from the English alphabet. Can we claim that letters in the set are members of the English alphabet?

Answer: Yes, with a probability of 5/5 = 1 or with 100% certainty.

Question 3: Assume that the letters are drawn from the English alphabet and another alphabet. Can we claim that letters in the set are members of the Chinese alphabet?

Answer: No, with a probability of 0/5 = 0 it is impossible.

Note: Odds in favor of X = probability of X divided by the probability of all other events. Odds against X = inverse of odds in favor of X.

Question 4: What are the odds in favor of R and the odds against K?

Answer: Odds in favor of R = (2/7)/[1-(2/7)] = 0.386
       Odds against R = 2.591.

Depending on the nature of an event and various philosophical conceptions of uncertainty, through time, several definitions of probability have been developed. Generally, they are referred to as the *classical, empirical, subjective*, and *axiomatic* definitions of probability. The remainder of the chapter borrows from and synthesizes the conceptual analysis found in the prominent works of Mosteller et al (1995), Hogg et al (1983), Ross (1988), Brunk (1975), Rosner (1995), Strait (1983) and Motulsky (1995.)

1.2.2.   Classical Probability

Classical probability concerns games with a finite number of equally likely results. If the event is the occurrence of any one of n possible results among the total number N of all possible results, then the probability of that event is defined to be n/N. For example, in rolling a conventional six-sided die, the probability of obtaining an odd number on the die is 3/6.

An important contribution made by classical probabilists is that the relative frequency ratio of a given event of certain games of chance tends to converge to a definite value when the game is repeated a great number of times. In games of chance, events that can be objectively determined to be equally likely are, in fact, observed to occur with equal frequency when the game is repeated many times. Thus, according to classical theory, they are events with equal probabilities. For example, the probability of 5 (or any number from 1 to 6) in rolling a die approaches 1/6 as the number of throws increases; similarly, the probability of heads (or tails) in tossing a coin approaches 1/2 as the number of tosses increases.

The classical theory can be extended beyond games of chance to any chance situation involving a finite number of equally likely outcomes. There are many problems, applied and theoretical, that fall into this category. Especially interesting and challenging problems of this sort are connected with combinatorial methods. Consider the following principles, definitions and some corresponding examples:

A. Multiplication Principle:

Let $X_1, X_2, \ldots X_k$ be sets consisting of $n_1, n_2, \ldots, n_k$ elements, respectively. There are $n_1.n_2\ldots n_k$ (where "." means times) ways to select first an element from X_1, then an element from X_2, then an element from $X_3, \ldots$, and finally an element from X_k.

Example #1:
A box contains 6 flat non-identical stones and 8 round non-identical stones. (a) Let a pair of stones consist of a flat stone and a round stone: what is the sample space or the number of pairs possible? (b) If there are 3 blue stones in A and 4 blue stones in B, (i) What is the probability that a random selection of stones will result in a pair of 2 blue stones? (ii) What is the probability that a random selection of stones will result in a pair of 2 non-blue stones? (iii) What is the probabil-

ity that a random selection of stones will result in a pair of 2 stones one of which is blue?

Answers to #1:
(a) The sample space consists of 48 sample points or N = 48.

(b) (i) Let E_{BB} = event "pairs of blue stones." Therefore, the number of points in event E_{BB} is N_{BB} = 12.
Hence, $P(E_{BB}) = N_{BB}/N = 12/48 = 1/4$.

(ii) Let E_{WW} = event "pairs of non-blue stones." Therefore, the number of points in event E_{WW} is N_{WW} = 12.
Hence, $P(E_{WW}) = N_{WW}/N = 12/48 = 1/4$.

(iii) Let E_{BW} = event "pairs of one blue and one non-blue stones." Therefore, the number of points in event E_{BW} is N_{BW} = 24.
Hence, $P(E_{BW}) = N_{BW}/N = 24/48 = 1/2$.

Example #2:
Your cousin George has 4 jackets, 8 shirts, 5 ties and 6 pairs of slacks.

(a) How many different outfits of a jacket, a shirt, a tie, and a pair of slacks are possible for him?

(b) If 2 jackets, 4 shirts, 3 ties and 3 pairs of slacks are brown, what is the probability that a random selection will result in an all-brown outfit?

Answers to #2:
(a) 4.8.5.6 = 960 different possible outfits; N = 960.

(b) 2.4.3.3 = 72 different all-brown outfits. Hence, if B = event of a completely brown outfit then, N_B = 7 and

$$P(B) = N_B/N = 72/960 = 0.075.$$

B. Sampling with Replacement:

If r objects are selected from a set of n objects, and if the order of selection is noted, the selected set of r objects is called an ordered sample of size r. Sampling with replacement occurs when an object is selected and then replaced before the next object is selected. When sample with replacement, the number of possible ordered samples of size r taken from a set of n objects is n^r.

Example #3:

(a) A coin is tossed 4 times. What is the number of possible ordered samples?

(b) A dodecahedron is rolled 3 times. What is the number of possible ordered samples?

(c) What is the numbers of possible four-letter code words using the four letters M, N, P, and Q?

Answers to #3:

(a) $2^4 = 16$

(b) $12^3 = 1,728$

(c) $4^4 = 256$

C. Permutations:

Suppose that n positions are to be filled with n different objects. Hence, there are n choices for filling the first position, n-1 for the second, n-2 for the third, . . . , 1 choice for the last position and, by the multiplication principle there are $n(n-1)(n-2) . . . (1)$ possible arrangements. Let all possible arrangements be equal to n! (the symbol n! is read n factorial) or,

$$n! = n(n-1)(n-2) . . . 2.1$$

Each of the n! arrangements (in a row) of n objects is called a permutation of the n objects and 0!=1 because zero positions can be filled with zero objects. If only r positions are to be filled with objects selected from n different objects, r≤n, the number of possible ordered arrangements is

$$P(n,r) = n(n-1)(n-2)...(n-r+1) = (n!) / [(n-r)!]$$

Each of the P(n,r) arrangements is called a permutation of n objects taken r at a time.

Example #4:

(a) What is 7!?

(b) What is R = (7! / 5!)?

(c) What is the number of permutations of the five letters R, M, C, T, and H?

(d) Based on the English alphabet (26 letters), what is the number of possible five-letter code words in which all five letters are different?

Answers to #4:
(a) 7.6.5.4.3.2.1 = 5,040
(b) R = (7.6.5!) / (5!) = 42
(c) 5! = 120
(d) P(26,5) = (26! / 21!) = (26.25.24.23.22.21!) / (21!) = 7,893,600

D. Sampling without Replacement:

Sampling without replacement occurs when an object is not replaced after it has been selected. Because P(n,r) represents the number of different ways that a group of r items could be selected from n items when the order of selection is relevant, and, as each group of r items will be counted r! times in the count, it follows that the number of different groups of r items that could be formed from a set of n items is [P(n,r)/r!] = (n!)/[(n-r)!r!].

Example #5:
(a) Pairs of 2 are to be formed from a group of 5 items without replacement. How many possible pairs are possible?

(b) Triplets of 3 are to be formed from a group of 7 items without replacement. How many triplets are possible?

Answers to #5:
(a) n = 5, r = 2; therefore, (5.4)/(2.1) = 10

(b) 35

1.2.3. Empirical Probability

Classical probability requires a finite number of equally likely results (the theory's major limitation) and as such it excludes the consideration of many chance phenomena of interest that may originate from unequally likely, continuous, dynamic, very large and infinite environments. As observers of natural phenomena may testify, the long-run stableness of relative frequency ratios is not restricted to games of chance; it may also be observed in various demographic, natural phenomena, historic and experimental data sets. How else may one explain the plausible probabilities estimated by actuaries in insurance companies or those estimated by meteorologists? For example, what does the meteorologist mean when she reports that the probability of a heat wave around this time of the

year (July) in this region is about 80%? Her conclusion is the result of many historic observations of July months in this region, in 80% of which we experienced
heat wave conditions.

One may define the empirical probability as follows:

$$P(X) = \lim_{n \to \infty} (n_X / n)$$

where P(X) = probability of event X, n = number of repetitions of an experiment,
n_X = number of times X occurs.

In practice, we accept the relative frequency ratio (n_X/n) as the P(X) if n is a
large number; in reality, it is not possible to repeat an experiment an infinite
number of times.

Consider an example found in Hogg et al (p. 4). Let the sample space (S) be
S = {1,2,3,4,5,6} and event (X) be X = {1,2}. The experiment consists of draw-
ing two numbers from S, n times, via computer simulation. What is the P(X)?
The results of the simulation in the following table indicate that the P(X) →
(1/3) as n → ∞.

N	n_X = Frequency	P(X) = N_X/n
50	16	0.32
100	34	0.34
250	80	0.32
500	163	0.326

But, although the empirical definition of probability is the most widely accepted definition of probability, the one that comes to mind most frequently, it is
problematical because it lacks rigor. One can always question the 'infinity' re-
quirement or whether or not n is large enough or how well the experiments are
executed. In general though, if the probability of an event X is P(X), then
$\lim_{n \to \infty}(n_X/n)$ *equals* P(X), if not always, then at least with the greatest likelihood.

Despite its limitations, in our opinion, empirical probability, has contributed
to the best-known result in probability theory, namely, the *strong law of large
numbers*. The law states that the average of a sequence of independent and iden-
tically distributed random variables will, with probability 1, converge to the
mean of the distribution. In other words,

$$[(X_1+X_2+...+X_n)/n] \to \mu \quad \text{as} \quad n \to \infty$$

or, $P\{\lim_{n \to \infty} (X_1+X_2+...+X_n)/n = \mu\} = 1$

1.2.4.   Subjective Probability

So far, probability of event X, P(X), has been defined as a long-run frequency of occurrence or,

$$P(X) = \lim_{n \to \infty} (n_X / n)$$

where n = number of repetitions of an experiment and n_X = number of times X occurs. But, often times we use the term probability to denote subjective belief. For example, if your opponent's backhand is weaker than her forehand, then it is more likely that you win a point when you play her backhand than when you play her forehand. In your judgement, event A (winning point off your opponent's backhand) is more likely to occur than event B (winning point off your opponent's forehand); as a result, you will assign a higher probability to A than to B such as P(A) = 0.6 and P(B) = 0.4 or P(A) = 0.7 and P(B) = 0.3.

Ideally, a person's subjective probabilities should be equal to 1 or, in general, consistent with the axioms of probability (see next section.) Naturally, of course, subjective probabilities are controversial.

Example #6:
Consider a 10-dog race for which you believe that each of the first 3 dogs has a 15% chance of winning, dogs 4 to 6 each has 10% chance, and the remaining 4 dogs, a 6.25% chance. Would it be better for you to bet that the winner will be one of the first four dogs, or to bet that the winner will be one of the dogs 1, 4, 6, 8, 10?

Answer to #6:
Based on your subjective probabilities, the first bet would be more attractive;

P(1,2,3,4 winning) = 0.55 > P(1,4,6,8,10 winning) = 0.475.

Example #7:
In the above example, if your subjective probability of the last dog winning changes from 0.0625 to 0.10, would you be able to compute the probabilities of the respective bets?

Answer to #7:
No, because the summation of your subjective probabilities exceeds 1.

1.2.5.   Axiomatic Probability

Definitions:

i. A statement that is assumed to be reasonable is called an *axiom.*

ii. A statement that can be deduced either from axioms or from previously proved theorems is called a *theorem* or a *proposition.*

iii. An *experiment* is a set of trials the results of which cannot be forecasted with certainty.

iv. The set of all possible outcomes of an experiment is called *sample space* (S.)

v. Each object in S is called an *element* of S.

vi. Any subset of S (or, collection of elements in S) is called an *event.*
Let {} = "the event."

Example #8:
Consider the experiment of rolling once, two symmetric (in all dimensions) tri-angular pyramids i and j (each pyramid has four faces numbered 1, 2, 3, 4.)

a.   The sample space of this experiment is:

S = {(i,j) i, j =1,2,3,4}			
1, 1	2, 1	3, 1	4, 1
1, 2	2, 2	3, 2	4, 2
1, 3	2, 3	3, 2	4, 3
1, 4	2, 4	3, 3	4, 4

b.   Pairs of numbers in the cells, such as (1, 1) or (4, 3) are elements, 16 all together.

c.   Event A may be defined as:
A = {(i,j): the i value is equal to the j value} = {(1,1), (2,2), (3,3), (4,4)};

d.   Event B may be defined as:
B = {(i,j): $i+j = 4$} = {(1,3), (2,2), (3,1)}.

vii. {A $\cup$ B} = the event that either A or B occurs or they both occur (see Venn Diagram 1.)

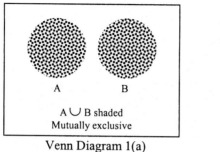

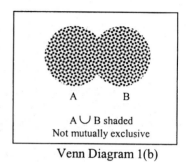

A ∪ B shaded
Mutually exclusive

Venn Diagram 1(a)

A ∪ B shaded
Not mutually exclusive

Venn Diagram 1(b)

viii. $\{A \cap B\}$ = the event that both A or B occur simultaneously (see Venn Diagram 2.)

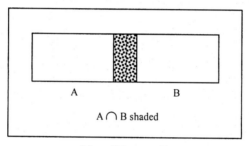

A ∩ B shaded

Venn Diagram 2

ix. $\overline{A}$ = the event that A does not occur (complement of A.) Because $\overline{A}$ occurs when A does not, $P(\overline{A}) = 1 - P(A)$. Events A and $\overline{A}$ are depicted in Venn Diagram 3.

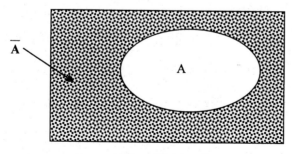

Venn Diagram 3

x. A and B are *independent events* if and only if $P(A \cap B) = P(A)P(B)$. Otherwise, if $P(A \cap B) \neq P(A)P(B)$, A and B are *dependent events*.

Example #9:
Let:
A = {mother's blood pressure ≥ 95} and $P(A) = 0.2$;
B = {father's blood pressure ≥ 95} and $P(B) = 0.3$;
C = {child's pressure ≥ 95 (mothered by A)} and $P(C) = 0.3$;
$P(A \cap B) = 0.06$;
$P(A \cap C) = 0.08$.

Therefore, A and B are independent because
$P(A \cap B) = P(A)P(B) \Rightarrow 0.06 = (0.2)(0.3)$; and
A and C are dependent because
$P(A \cap C) \neq P(A)P(C) \Rightarrow 0.08 \neq (0.2)(0.3)$.

Example #10:
Consider the sample space in Example #8.

Let:
A = {1 or 2 on the second roll},
B = {2 or 3 on the first roll}.

Then,
$P(A) = 8/16$, $P(B) = 8/16$, $P(A \cap B) = 4/16$ and
$P(A)P(B) = (8/16)(8/16) = 4/16 = P(A \cap B)$.
Hence A and B are independent events.

xi. The above definition can be generalized to the case of m independent events which is referred to as *the multiplication law of probability* (stated above in the Classical Probability section.) Hence, if $A_1, \ldots A_m$ are mutually independent events, then

$$P(A_1 \cap A_2 \cap \ldots \cap A_m) = P(A_1)P(A_2) \ldots P(A_m).$$

Axiom I: A probability is a number between, and inclusive of, 0 and 1. In other words, the probability of an event A, denoted by $P(A)$, always satisfies

$$0 \leq P(A) \leq 1.$$

Axiom II: In performing an experiment, we assume that the result will be one of the elements. With probability 1, the outcome will be an element in the sample space S, or

$$P(S) = 1.$$

Axiom III: If A_1, A_2, A_3, . . . are events that cannot happen simultaneously, then the probability that one of them will occur is the sum of their probabilities. For example, the probability of either 1, 2 or 3 in a roll of a die is $1/6 + 1/6 + 1/6 = 3/6$. Hence, if A_1, A_2, A_3, . . . is a finite or infinite sequence of mutually exclusive events of S, and $A_i \cap A_j = \varnothing$ (where $\varnothing$ is the empty or null set), $i \neq j$, then

$$P(A_1 \cup A_2 \cup A_3 \cup \ldots) = P(A_1) + P(A_2) + P(A_3) + \ldots$$

Based on the above axioms, one may derive the following theorems:

Theorem I: $P(A) = 1 - P(\overline{A})$.

Notice: $S = A \cup \overline{A}$;

hence, applying from axioms II and III we have

$$P(S) = P(A) + P(\overline{A}) \Rightarrow$$

$$1 = P(A) + P(\overline{A}) \Rightarrow$$

$$P(A) = 1 - P(\overline{A}).$$

Theorem II: $P(\varnothing) = 0$.

Notice: $\varnothing \cup S = S$; hence,

$$P(\varnothing) + P(S) = P(S) \Rightarrow$$

$$P(\varnothing) = P(S) - P(S) = 0.$$

Theorem III: Addition Law for any two events or,

$$P(A \cup B) = P(A) + P(B) - P(A \cap B).$$

Consider Venn Diagram 4 in conjunction with sets A, B and three mutually exclusive regions x, y and z:

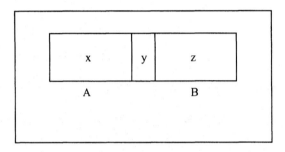

Venn Diagram 4

$A = x \cup y$;

$B = y \cup z$;

$A \cup B = x \cup y \cup z$;

from Axiom III,

$P(A) = P(x) + P(y)$;

$P(B) = P(y) + P(z)$;

$P(A \cup B) = P(x) + P(y) + P(z)$;
add and subtract $P(y)$ in the last equation so that

$P(A \cup B) = P(x) + P(y) + P(z) + P(y) - P(y)$; hence,

$P(A \cup B) = P(A) + P(B) - P(A \cap B)$.

The addition law may be extended to more than two events. For any three events A, B, C it may be expressed as follows:

$$P(A \cup B \cup C) = P(A) + P(B) + P(C)$$

$$- P(A \cap B) - P(A \cap C) - P(B \cap C)$$

$$+ P(A \cap B \cap C).$$

Example #11:
Consider S = {(i,j) i, j =1,2,3,4} the sample space of Example #10.

Let:
A = {(i,j): the i value is equal to the j value}
   = {(1,1), (2,2), (3,3), (4,4)} and

B = {(i,j): i + j = 4}
   = {(1,3), (2,2), (3,1)}.

Clearly, A ∩ B = {(2,2)};

then P(A) = 4/16,
    P(B) = 3/16,
    P(A ∩ B) = 1/16, and
    P(A ∪ B) = 4/16 + 3/16 – 1/16 = 6/16.

Example #12:
If the respective probabilities for events A, B, C are:

$$P(A) = 1/2, P(B) = 1/3 \text{ and } P(C) = 1/4$$

then, given the intersections

$$A \cap C = \varnothing, B \cap C = \varnothing \text{ and } P(A \cap B) = 1/6,$$

the following may be computed:

a. $P[\overline{(A \cap B)}] = 1\text{-}P(A \cap B) = 1 - 1/6 = 5/6.$

b. $P(A \cap \overline{B}) = 1/3$ because
    $A = (A \cap \overline{B}) \cup (A \cap B)$ and
    $P(A) = P[(A \cap \overline{B}) \cup (A \cap B)]$
    $= P(A \cap \overline{B}) + P(A \cap B) \Rightarrow$
    $1/2 = P(A \cap \overline{B}) + 1/6 \Rightarrow$
    $P(A \cap \overline{B}) = 1/3.$

c. $P[\overline{(A \cup B)}] = 1/3$ because

$P[\overline{(A \cup B)}] = 1 - P(A \cup B)$ where
$P(A \cup B) = P(A) + P(B) - P(A \cap B) = 1/2 + 1/3 - 1/6 = 2/3;$
therefore, $P[\overline{(A \cup B)}] = 1 - 2/3 = 1/3.$

d. $P(\overline{A} \cap \overline{B}) = 1/3$ because, as per de Morgan Law,
$(\overline{A} \cap \overline{B}) = \overline{(A \cup B)}$; therefore,
$P(\overline{A} \cap \overline{B}) = P[\overline{(A \cup B)}] = 1/3.$

Theorem IV: Addition Law for Independent events. If two events A
and B are independent, then $P(A \cap B) = P(A)(P(B)$ and

$$P(A \cup B) = P(A) + P(B) - P(A \cap B) \quad \Rightarrow$$

$$P(A \cup B) = P(A) + P(B) - P(A)(P(B) \Rightarrow$$

$$P(A \cup B) = P(A) + P(B)[1 - P(A)].$$

Example #13:
The following two events are independent:

$A = \{$mother's blood pressure $\geq 95\}$ with $P(A) = 0.2;$
$B = \{$father's blood pressure $\geq 95\}$ with $P(B) = 0.3.$

Based on these two probabilities one may compute the probability of a hyperten-
sive household as follows:

$$P(A \cup B) = P(A) + P(B)[1 - P(A)] = (0.2) + (0.3)(0.8) = 0.44.$$

1.2.6.   Conditional Probability

Conditional probability is an essential and practical concept especially when
we want to compute the probability of several non-independent events. As the
name suggests, the conditional probability of an event is the probability that an
event will occur, given the probability of another event.

More formally, the conditional probability of an event A, given that event B
has occurred with $P(B) > 0$, is defined by

$$P(A/B) \quad = \quad \frac{P(A \cap B)}{P(B)} \, ,$$

where P(A/B) is read as the probability of A given that B has occurred.

Similarly, the conditional probability of an event B, given that event A has occurred with P(A) > 0, is defined by

$$P(B/A) \quad = \quad \frac{P(B \cap A)}{P(A)} \, ,$$

where P(B/A) is read as the probability of B given that A has occurred.

Note: For convenience purposes, in some of the following sections, the term $P(A_i \cap A_j)$ will be denoted by $P(A_iA_j)$ and the term $\bar{A}$ by A'.

Example #14:
The following table reports screening results for a test devised for a certain disease.

	Disease Present(H)	Disease Absent (I)	Total
Positive Test (F)	35	165	200
Negative Test (G)	20	280	300
Total	55	445	500

a.   The probability of a positive test given the probability that the disease is present is:

$$P(F/H) = \frac{P(F \cap H)}{P(H)} = \frac{35/500}{55/500} = \frac{35}{55} \, .$$

b.   The probability of a positive test given the probability that the disease is absent is:

$$P(F/I) = \frac{P(F \cap I)}{P(I)} = \frac{165/500}{445/500} = \frac{165}{445} \, .$$

c.   The probability of a negative test given the probability that the disease is absent is:

$$P(G \mid I) = \frac{P(G \cap I)}{P(I)} = \frac{280/500}{445/500} = \frac{280}{445} \ .$$

Theorem I: Based on the definition for conditional probability for two events, by the multiplication rule,

$$P(A \cap B) = P(A)P(B/A) \text{ and } P(B \cap A) = P(B)P(A/B);$$

for m events, the *Generalized Multiplication Law of Probability* for a set of m arbitrary events is

$$P(A_1A_2 \ldots A_m) = P(A_1)P(A_2/A_1)P(A_3/A_1A_2) \ldots$$
$$P(A_m/A_1A_2 \ldots A_{m-1}).$$

Theorem II:

$$P(A'/B) = 1 - P(A/B).$$

Notice that $P(A'B) = P(B) - P(AB)$; therefore,

$$P(A'/B) = P(A'B) / P(B) =$$
$$[P(B) - P(AB)] / P(B) =$$
$$[P(B)/P(B)] - [P(AB)/P(B)] =$$
$$1 - P(A/B).$$

Example #15:
60% of the population of adults in a country consists of college graduates; 80% of them have incomes over $50,000. (a) What is the probability that a randomly selected college graduate has an income of $50,000 or less? (b) What percent of adults are college graduates and have incomes over $50,000?

Answers to #15:
Let:
A = {adult with income over $50,000};
B = {college graduate adult};

Hence, $P(A/B) = 0.8$ and $P(B) = 0.6$.

Therefore,
(a) $P(A'/B) = 1 - P(A/B) = 1 - 0.8 = 0.2$ and

(b) $P(AB) = P(A/B)P(B) = (0.8)(0.6) = 0.48$.

Example #16:
Three good computer disks have been mixed up with two defective ones. The disks are tested, one by one, until both defective ones are found.

(a) What is the probability that the testing of exactly two disks is required?

(b) What is the probability that the testing of exactly three disks is required?

Answers to #16:
Let $i = 1,2,3,4,5$; $G_i = \{$good disk on the ith test$\}$ and $D_i = \{$defective disk on the ith test$\}$.

(a) The event that the testing of two disks is required is D_1D_2 with a probability of $P(D_1D_2) = P(D_1)P(D_2/D_1)$. Obviously, $P(D_1) = 2/5$ (because in choosing a disk for the first test there are 5 disks of which 2 are defective) and $P(D_2/D_1) = 1/4$ (because if the first disk tested is defective, there remain 4 disks of which 1 is defective.) Therefore,
$P(D_1D_2) = P(D_1)P(D_2/D_1) = (2/5)(1/4) = 0.10$

(b) The event that the testing of 3 disks is required is
$\{G_1D_2D_3 \cup D_1G_2D_3 \cup G_1G_2G_3\}$ (if the first 3 disks tested are good, the remaining 2 must be defective).

Hence,
$P(G_1D_2D_3 \cup D_1G_2D_3 \cup G_1G_2G_3) =$

$P(G_1D_2D_3) + P(D_1G_2D_3) + P(G_1G_2G_3) =$

$P(G_1)P(D_2/G_1)P(D_3/G_1D_2) + P(D_1)P(G_2/D_1)P(D_3/D_1G_2) +$
$P(G_1)P(G_2/G_1)P(G_3/G_1G_2) =$

$(3/5)(2/4)(1/3) + (2/5)(3/4)(1/3) + (3/5)(2/4)(1/3) = 0.3$.

Theorem III: Total Probability Rule

If $A_1, A_2, \ldots A_m$ are mutually exclusive and exhaustive events in S such that

$S = A_1 \cup A_2 \cup \ldots \cup A_m$ and

$P(A_i) > 0$ for $i = 1, 2, \ldots, m$.

Then for any event B of S

$$P(B) = P(BA_1) + P(BA_2) + \ldots + P(BA_m) \text{ or}$$

$$P(B) = \sum_{i=1}^{m} P(B/A_i)P(A_i).$$

Venn Diagram 5 illustrates the conditions for the theorem, given a case of m = 4. B occurs together with one and only one of the events A events; B is the union of the mutually exclusive and exhaustive events BA_1, BA_2, BA_3, and BA_4. Therefore, $P(B) = P(BA_1) + P(BA_2) + P(BA_3) + P(BA_4)$.

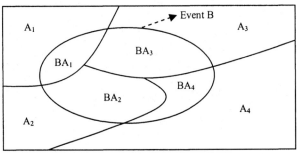

Venn Diagram 5

Example #17:
The population of airplane pilots in a country is 5,000 men and women 50 years and older. We know from census data that 45% of the population is age 50-54, 28% is age 55-59, 20% is age 60-64, and 7% is age 65 or older. We also know from previous studies that 2.4%, 4.6%, 8.8%, and 15.3% of the men and women in those respective age groups will develop eye problems. What percentage of the population will develop eye problems, and how many cases does this percentage represent?

Answers to #17:
Let $A_i = \{\text{Age Group}_i\}$ where i = 1, . . . 4; hence,
$P(A_1) = 0.45, P(A_2) = 0.28, P(A_3) = 0.2, P(A_4) = 0.07$ and
$P(B/A_1) = 0.024, P(B/A_2) = 0.046, P(B/A_3) = 0.088, P(B/A_4) = 0.153$; and

$$P(B) = \sum_{i=1}^{m=4} P(\frac{B}{A_i})P(A_i) = 0.052.$$

Therefore, 5.2% of the population will develop eye problems, which represents a total of $(5,000)(0.052) = 260$ cases.

Theorem IV: Bayes' Rule (Bayesian Theorem)

If A is a *hypothesis* and B is *given evidence*, in its simplest form, provided that $P(B) > 0$, the theorem is

$$P(A/B) = \frac{P(B \cap A)}{P(B)} = \frac{P(\frac{B}{A})P(A)}{P(B)} ,$$

where P(A/B) = *posterior probability* of A, and
P(A) = *prior probability* of A before collecting B.

Generally, if the conditions of the above theorem are satisfied, then for any $k = 1, 2, \ldots m$

$$P(A_k / B) = \frac{P(\frac{B}{A_k})P(A_k)}{\sum_{i=1}^{m}(\frac{B}{A_i})P(A_i)}$$

In other words, the Bayesian formula is the ratio of one of the BA areas in Venn Diagram 5 over the sum of all BA areas.

Example #18:
Consider the following table, which reports production data by 3 machines producing X in a factory:

	Production percentages	Defective percentages (prior probabilities)
Machine A_1	35%	2%
Machine A_2	25%	1%
Machine A_3	40%	3%

(a) If one X is selected at random from among all X's produced, what is the probability that it is defective (B)?

(b) If the selected X is defective, what is the probability that it was produced by
A_1, A_2, A_3? Compare the prior to the posterior probabilities.

Answers to #18:

(a) $P(B) = \sum_{i=1}^{3} P\left(\dfrac{B}{A_i}\right) P(A_i) =$

   $P(B/A_1)P(A_1) + P(B/A_2)P(A_2) + P(B/A_3)P(A_3) =$

   $(0.35)(0.02) + (0.25)(0.01) + (0.40)(0.03) = 0.0215$

(b) $P(A_1 / B) = \dfrac{P\left(\dfrac{B}{A_1}\right) P(A_1)}{P(B)} = \dfrac{(0.02)(0.35)}{0.0215} = 0.3255$

   $P(A_2 / B) = \dfrac{P\left(\dfrac{B}{A_2}\right) P(A_2)}{P(B)} = \dfrac{(0.01)(0.25)}{0.0215} = 0.1163$

   $P(A_3 / B) = \dfrac{P\left(\dfrac{B}{A_3}\right) P(A_3)}{P(B)} = \dfrac{(0.03)(0.40)}{0.0215} = 0.5581$

All posterior probabilities are higher than the prior probabilities.

1.3 Screening Tests, Epidemiological Tests and Bayesian Analysis

   Suppose a disease-type population y screening test, has generated the results
in the following table where upper case letters correspond to events and lower
case letters to numbers. Venn Diagrams 6 and 7 display the events in the following table:

	Disease Present (H)	Disease Absent (I)	Total
Positive Test (F)	K	R	S
Negative Test (G)	T	U	V
Total	W	X	Y

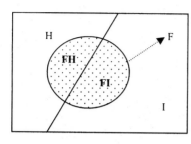 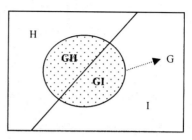

Venn Diagram 6   Venn Diagram 7

Conventionally, the following definitions, and their modifications in conjunction with the previous table and Venn Diagrams 6 and 7, apply:

i. Positive Prior Probability or Prevalence = $P(H) = W / Y$.

Note: In a *population study*, the positive prior probability is called prevalence. Additionally, a positive prior probability may be obtained from *genetic theory* or *clinical intuition*.

ii. Negative Prior Probability = $P(H) = X / Y$.

iii. Sensitivity = $P(F/H) = K / W$.

iv. Specificity = $P(G/I) = U / X$

v. PV $^+$ = Positive Predictive Value = $P(H/F) = K / S$ where

$$P(H / F) = \frac{P(F / H)P(H)}{P(F / H)P(H) + P(F / I)P(I)} ;$$

because $P(F/I) + P(G/I) = 1 \Rightarrow$
$P(F/I) = 1 - P(G/I) = 1 - $ Specificity,

and $P(I) + P(H) = 1 \Rightarrow$
$P(I) = 1 - P(H),$

we have,

$$PV^+ = \frac{(Sensitivity)(Prevalence)}{(Sensitivity)(Prevalence) + (1 - Specificity)(1 - Prevalence)}$$

vi. PV $^-$ = Negative Predictive Value = P(I/G) = U / V; similarly,

$$PV^- = \frac{(Specificity)(1-Prevalence)}{(Specificity)(1-Prevalence)+(1-Sensitivity)(Prevalence)}$$

vii. False$^+$ = False Positives = P(I/F) = R / S; or,

$$P(I/F) = \frac{P(F/I)P(I)}{P(F/I)P(I)+P(F/H)P(H)} \text{ ; and}$$

viii. False$^-$ = False Negatives = P(H/G) = T / V;

$$P(H/G) = \frac{P(G/H)P(H)}{P(G/H)P(H)+P(G/I)P(I)} \cdot$$

The Bayesian Theorem provides a way for considering two hypotheses, for example H and I, in terms of odds ratios.

Note: Odds in favor of H is expressed as follows:

$$Odds_H = \frac{P(H)}{1-P(H)} \qquad \Rightarrow \qquad P(H) = \frac{Odds_H}{1+Odds_H} \cdot$$

Likewise, because

$$P(H/F) + P(I/F) = 1 \Rightarrow P(I/F) = 1 - P(H/F)$$

where, $P(H/F) = \dfrac{P(F/H)P(H)}{P(F/H)P(H)+P(F/I)P(I)}$

and, $P(I/F) = \dfrac{P(F/I)P(I)}{P(F/I)P(I)+P(F/H)P(H)}$; then,

the Odds$_{P(H/F)}$ = Positive Posterior Odds = $\dfrac{P(H/F)}{1-P(H/F)} = \dfrac{P(H/F)}{P(I/F)}$ and, since the denominators cancel out,

Positive Posterior Odds = $\dfrac{P(F/H)}{P(F/I)} \dfrac{P(H)}{P(I)} = \dfrac{(Sensitivity)}{(1-Specificity)} \dfrac{(Prevalence)}{(1-Prevalence)}$

where, $\dfrac{(Sensitvity)}{(1-Specificity)}$ = Positive Likelihood,

$\dfrac{(Prevalence)}{(1-Prevalence)}$ = Positive Prior Odds, and

Positive Posterior Probability = $\dfrac{Positive \;\; Posterior \;\; Odds}{1+Positive \;\; Posterior \;\; Odds}$.

Note: Positive Posterior Probability = PV $^{+}$.

Similarly,

Negative Posterior Odds = $\dfrac{P(G/I)}{P(G/H)} \dfrac{P(I)}{P(H)} = \dfrac{(Specificity)}{(1-Sensitivity)} \dfrac{(1-Prevalence)}{(Prevalence)}$

where, $\dfrac{(Specificity)}{(1-Sensitivity)}$ = Negative Likelihood,

$\dfrac{(1-Prevalence)}{(Prevalence)}$ = Negative Prior Odds, and

Negative Posterior Probability = $\dfrac{Negative \;\; Posterior \;\; Odds}{1+Negative \;\; Posterior \;\; Odds}$.

Note: Negative Posterior Probability = PV $^{-}$.

Note: For a *fatal disease*, it pays to set the *sensitivity very high* even at the expense of low specificity; on the other hand, for an *incurable non-communicable disease*, it pays to set the *specificity very high* even at the expense of low sensitivity. (For more details on this, see Motulsky, Chapter 14.)

Example #19:

PART A: Population-Screening Test Results

The following table shows the results of a 1,000,000 people population-screening test regarding a disease.

	Disease Present (H)	Disease Absent (I)	Total
Positive Test (F)	90	37,000	37,090
Negative Test (G)	20	962,890	962,910
Total	110	999,890	1,000,000

Prevalence (or, Positive Prior Probability) = 0.00011;
Sensitivity = 0.818;
Specificity = 0.963;
Positive Prior Odds = (0.00011) / (1 – 0.00011) = 0.00011;
Positive Likelihood = (0.818) / (1 – 0.963) = 22.11;
Positive Posterior Odds = (22.11)(0.00011) = 0.0024;
Positive Posterior Probability = (0.0024) / (1 + 0.0024) = 0.00239.

Hence, if a person is diagnosed with the disease, the probability that the test is accurate is 0.00239 whereas the probability that it is inaccurate, or the $P(False^+)$, is 0.998.

Similarly,
Negative Prior Probability = 0.999;
Negative Prior Odds = (1 - 0.00011) / (0.00011) = 9,089.91
Negative Likelihood = (0.963) / (1 - 0.818) = 5.29;
Negative Posterior Odds = (5.29)(9,089.91) = 48,081.34;
Negative Posterior Probability = (48,081.34) / (1 + 48,081.34) = 0.999.

Hence, if a person is tested and declared free of the disease, the probability that the test is accurate is 0.999 whereas the probability that it is inaccurate, or the $P(False^-)$, is 0.001.

A disease may affect a subgroup of the screened population differently than another. Consider below Part B of Example #19 followed by a summary.

PART B: Epidemiological Test Results based on Women

Suppose, genetic theory has predetermined that there is 50.5% chance that a randomly selected woman from the population carries the gene. 1,000 women were randomly selected from the above population and tested for the disease under consideration. Therefore, given the population-screening test's values for Sensitivity = 0.818 and Specificity = 0.963 derived in Part A above:

Positive Prior Probability for Women = 0.505;

Positive Posterior Odds for Women = (Positive Likelihood)(Positive Prior Odds)
= (22.11)(1.02) = 22.556;
Positive Posterior Probability for Women = 0.96.

Hence, if a woman is diagnosed with the disease, the probability that the test is accurate is 0.96 whereas the probability that it is inaccurate, or the P(False$^+$), is 0.04.

Similarly,
Negative Prior Probability for Women = (1,000-505) / (1,000) = 0.495;
Negative Posterior Odds for Women = (Negative Likelihood)(Negative Prior Odds) = (5.29)(0.98) = 5.185;
Negative Posterior Probability = 0.838.

Hence, if a woman is tested and declared free of the disease, the probability that the test is accurate is 0.838 whereas the probability that it is inaccurate, or the P(False$^-$), is 0.162.

Summary of Results for Example #19

	+ or -	Prior Prob.	Prior Odds	Posterior Odds	Posterior Prob.
Population	+	0.00011	0.00011	0.0024	0.00239
	-	0.999	9,089.91	48,081.34	0.999
Women	+	0.505	1.02	22.556	0.96
	-	0.495	0.98	5.185	0.838

Concluding, the following quotation about Bayesian analysis, regarding the interpretation of lab tests, may be helpful. According to Motulsky (p.147),

> When interpreting the result of a lab test for a particular patient, you must integrate what is known about the accuracy of the laboratory test (sensitivity and specificity) with what is known about the patient (prevalence, or prior probability that the patient has disease.) Bayesian logic combines these values precisely.

Example #20:

The application of "Bayesian" probability need not be limited to medical and health applications. Consider, for example, the release of a new movie. The producing company judges the probability of a "smashing" success is 0.3 (i.e. "prior" probability). It also knows that, of a total of 200 movies reviewed by a particular reviewer, the following results were obtained:

	Movie was Successful	Movie was Failure	Total
Reviewer Liked	75	10	85
Reviewer Disliked	25	90	115
Total	100	100	200

The movie company wishes to use the "Bayesian" approach in order to revise its probability-of-success figure: 1) if it learned that the reviewer praised the movie and; 2) if the reviewer criticized the movie.

Answers to #20:
1.   Reviewer Praised Movie:

The sensitivity measure indicates the movies the reviewer liked which were successful. Here, sensitivity = 75/100 = 0.75. In addition, the specificity measure indicates the movies the reviewer disliked which were failures. Here, specificity = 90/100 = 0.9. Lastly, "prior" odds = 0.3/0.7 = 0.43. Thus, applying the "Bayes Equation"

Posterior Odds = 0.43 [0.75 / (1-0.9)] = 3.225

From this, the "posterior" probability-of-success = 3.225 / (1+3.225) = 0.76

2.   Reviewer Criticized Movie:

The sensitivity measure indicates the movies the reviewer disliked which were successful. Here, sensitivity = 25/100 = 0.25. In addition, the specificity measure indicates the movies the reviewer liked which were failures. Here, specificity = 10/100 = 0.1. Lastly, prior odds = 0.3/0.7 = 0.43. Thus, applying the "Bayes" Equation

Posterior Odds = 0.43 [0.25 / (1-0.1)] = 0.119

From this, the "posterior" probability-of-success = 0.119 / (1+0.119) = 0.11

Chapter 2

PROBABILITY DISTRIBUTIONS, SUMMARY
MEASURES AND GRAPHS

*"The origin of graphs lies in the stars and planets
in the heavens above"*

2.1. Probability Distributions

A *probability distribution* is always associated with a man-made or natural experiment and it describes all possible outcomes and their corresponding likelihoods. In other words, it is a listing of the outcomes of an experiment that may occur and their corresponding probabilities.

A probability distribution is described by three sets of numbers: the possible outcomes of the experiment (or the sample space), the frequency of occurrence of each possible outcome, and the probability associated with each possible outcome. Each probability ranges in a closed interval from 0 to 1, and the summation of all probabilities that comprise a probability distribution equals 1.

It is important to note that, traditionally, textbooks that present material under the heading "Descriptive Statistics" would refer to this construct as a "frequency distribution" with the corresponding graphical representations as "histograms" and "polygons."

2.1.1 Basics

For example, suppose population X_1 of size $N = 10$ is represented by the following numerical values or data:

16, 41, 25, 21, 30, 17, 29, 50, 30, 39

X's possibilities (or sample space), their frequencies, and their probabilities are listed in Table 2.1:

Table 2.1

Sample Space (X₁)	Frequency (f)	Probability (P)
16	1	1/10
41	1	1/10
25	1	1/10
21	1	1/10
30	2	2/10
17	1	1/10
29	1	1/10
50	1	1/10
39	1	1/10
	10	1

The probability distribution described in Table 2.1 is shown graphically in Figure 2.1. P, expressed as a percent, is measured on the vertical axis, while X_1 on the horizontal. Graphs like Figure 2.1 are called *histograms* or *bar charts*.

Figure 2.1

Probability %

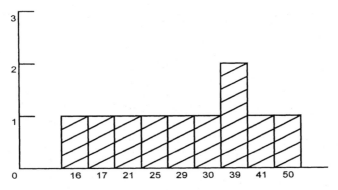

Let X be a random variable. A *Random variable* is a set of experimental outcomes each one of which is attached a probability of occurrence. If the possible values of a random variable are countable at specific points on a scale of values, as in 1, 5, 6, 9, 11, 34, and so forth, the variable is said to be *discrete*. If the possible values of a random variable can assume infinite values at all points on a scale of values, the variable is said to be *continuous*. Thus, depending on how a variable is described, probability distributions may be characterized as discrete or continuous.

Random variables may be described in an *ungrouped* fashion, as we did in Table 2.1 and Figure 2.1 above, and/or in a *grouped* fashion. Grouping of large data sets is helpful as it enables the researcher to see more clearly the data's general message. Grouping is more important the larger the data set gets: As sets of data become larger it becomes progressively more difficult to see the "forest" instead of the "trees." Large arrays of numerical observations are extremely difficult to inspect directly in order to get a notion of their general characteristics. Grouped descriptions of random variables are also more visually appealing. Casual inspection of a random variable's grouped histogram quickly reveals the general pattern of the underlying probability distribution.

The results of any experiment may be grouped in K > 0 "equal-width" intervals or classes as follows:

(a) The width (W) of every interval may be approximated by subtracting the lowest numerical value (L) from the highest (H) and dividing the difference by K. Or, W = (H-L)/K.

(b) For X_1, if K = 4, then W = (50-16)/4 = 8.5 or approximately 9.

Table 2.2 summarizes the results.

Table 2.2

Sample Space (X_1) Intervals	Frequency (f)	Probability (P)	Midpoint (m)
16 – 24	3	3/10	20
25 – 33	4	4/10	29
34 – 42	2	2/10	38
43 – 51	1	1/10	47
	10	1	

Starting with the lowest number, 16, all intervals are constructed on the basis of the same width, which in this case is 9. The first interval has a frequency of 3, because there are only three numbers that belong in this interval, a combined probability of 3/10, and a midpoint of (16+24)/2 = 20. The midpoint is the average of the interval's limits. Similarly for the rest of the classes. Figure 2.2 is the grouped histogram of variable X_1. Like in Figure 2.1, the vertical axis measures P and the horizontal X_1.

Figure 2.2

Probability

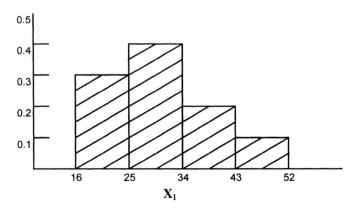

$$\mathbf{X_1}$$

   A random variable may be described by a variety of graphs. The most popular graph is the histogram (see Figures 2.1 and 2.2). Other graphs are the box-
and-whiskers plot, the stem-and-leaf display, the fuzzygram, the dot plot, the
ogive, the pie chart, etc. Box-and-Whiskers plots and Stem-and-Leaf displays are
becoming increasingly popular. Both of them will be discussed in section 2.1.4.
They will be better understood after a discussion of summary measures of central
tendency and variation.
   A random variable may also be described by summary measures like means,
percentiles, modes, the variance, the standard deviation, and the coefficient of
variation. These summary measures are called population parameters.

2.2 Population Parameters and Sample Statistics

   Population parameters are usually denoted by Greek letters.

We will use:

- μ (read as "mu") for *mean*,
- σ^2 (read as "sigma squared") for *variance*,
- σ (read as "sigma") for *standard deviation*,
- δ (read as "delta") for *percentile*,
- λ (read as "lamda") for *mode*,
- π (read as "pi") for *skewness*,
- θ (read as "theta") for *kurtosis*, and
- N   (read as capital "nu") for *population size*.

Sample statistics are usually denoted by Roman letters.

We will use:

- $\overline{X}$ (read X bar) for *sample mean,*
- s^2 for *sample variance,*
- s for *sample standard deviation,*
- M_d for *sample percentile,*
- M_o for *sample mode,*
- S_k for *sample skewness,*
- K for *sample kurtosis,* and
- n for *sample size*

Important Note: All statistics are computed exactly like the population parameters, except s^2 and s which are divided by (n-1). Dividing by (n-1), instead of n, makes s^2 and s more accurate estimators of the usually unknown population σ^2 and σ. For more on this, please see the discussion on "degrees of freedom" in Chapter 3.

As it has already been stressed, in inferential statistics the goal is to learn about a population of interest by relying on randomly drawn samples from this population. In general, the objective is to learn about certain population parameters by relying on sample statistics through inference. See Table 2.3.

Table 2.3 – Variable X

Sample Mean ($\overline{X}$)	$\rightarrow$	Inference	$\rightarrow$	Population Mean (μ)
Sample Variance (S^2)	$\rightarrow$	Inference	$\rightarrow$	Population Variance (σ^2)
Sample Percentile (M_d)	$\rightarrow$	Inference	$\rightarrow$	Population Percentile (δ)
Sample Statistic	$\rightarrow$	Inference	$\rightarrow$	Corresponding Population Parameter

2.3 Measures of Central Tendency and Variation

The primary measure of central tendency is the *Mean*. For the following discussion, let us consider the Mean of a discrete random variable X is a weighted average of its values, where the weights are the associated probabilities. It is also called expected value or E(X).

(At this point, the reader should review Appendix II: Summation Notation. In whatever follows, the indexing of N is omitted for convenience.)

For *ungrouped (un) data*, as in Table 2.1,

$$\mu_{un} = E(X) = \Sigma\, [\, X \cdot P(X)\,]$$

Example #1:
Consider Table 2.1.

$$\mu_{un} = E(X_1) = \Sigma[\, X_1 \cdot P(X_1)] = (16)(1/10) + (41)(1/10) + \ldots +$$
$$(39)(1/10) = 29.8$$

For *grouped (gr) data*, as in Table 2.2,

$$\mu_{gr} = \frac{\Sigma f \cdot m}{N}$$

Example #2:
Consider Table 2.2.

$$\mu_{gr} = \frac{\Sigma f \cdot m}{N} = \frac{(3)(20)+(4)(29)+(2)(38)+(1)(47)}{10} = \frac{299}{10} = 29.9$$

Notice that Tables 2.1 and 2.2 describe the same set of data but the numerical values for μ_{un} and μ_{gr} are not identical. This is due to grouping which, to a very small degree, distorts measures of central tendency.

Alternatively, μ may be computed from *raw (r) data* as follows:

$$\mu_r = \frac{\Sigma X}{N}$$

Example #3:
Assume the raw data for X_1 are 16, 41, 25, 21, 30, 17, 29, 50, 30, and 39. Therefore,

$$\mu_r = \frac{\Sigma X}{N} = \frac{(16+41+25+21+30+17+29+50+30+39)}{10} = \frac{298}{10} = 29.8$$

The primary measure of variation is called the *Variance*. For a discrete random variable X, variance measures how the X values fluctuate relative to their mean and is denoted by σ^2. It is a weighted average of the squared deviations of each X value from the mean; the weights are the probabilities associated with the values of X.

For *ungrouped data*, as in Table 2.1,

$$\sigma^2_{un} = \Sigma \, (X - \mu_{un})^2 \cdot P(X)$$

Example #4:
Consider Table 2.1 and the value for μ_{un} that we computed above.

$$\sigma^2_{un} = \Sigma \, (X - \mu_{un})^2 \cdot P(X)$$

$$= [(16 - 29.8)^2 \cdot (1/10)] + [(41 - 29.8)^2 \cdot (1/10)] + \ldots + [(39 - 29.8)^2 \cdot (1/10)]$$

$$= [(-13.8)^2 \cdot (0.1)] + [(11.2)^2 \cdot (0.1)] + \ldots + [(9.2)^2 \cdot (0.1)]$$

$$= [(190.44) \cdot (0.1)] + [(125.44) \cdot (0.1)] + \ldots + [(84.64) \cdot (0.1)]$$

$$= 19.044 + 12.544 + \ldots + 8.464 = 107.36$$

For *grouped data (gr)*, as in Table 2.2,

$$\sigma^2_{gr} = \frac{\Sigma f \cdot (m - \mu_{gr})^2}{N}$$

Example #5:
Consider Table 2.2 and the value for μ_{gr} that we computed above.

$$\sigma^2_{gr} = \frac{\Sigma f \cdot (m - \mu_{gr})^2}{N}$$

$$= \frac{[3 \cdot (20 - 29.9)^2 + 4 \cdot (29 - 29.9)^2 + 2 \cdot (38 - 29.9)^2 + 1 \cdot (47 - 29.9)^2]}{10}$$

$$= \frac{[3 \cdot (-9.9)^2 + 4 \cdot (-0.9)^2 + 2 \cdot (8.1)^2 + 1 \cdot (17.1)^2]}{10}$$

$$= \frac{[(3)(98.01) + (4)(0.81) + (2)(65.61) + (1)(292.41)]}{10}$$

$$= \frac{[294.03 + 3.24 + 131.22 + 292.41]}{10} = \frac{720.9}{10} = 72.09$$

Again, σ^2_{gr} is different than σ^2_{un} due to grouping.

Alternatively, σ may be computed from *raw data (r)* as follows:

$$\sigma^2_r = \frac{\Sigma(X - \mu_r)^2}{N}$$

Example #6:
Consider the raw data for X_1: 16, 41, 25, 21, 30, 17, 29, 50, 30, 39, and the value for μ_r that we computed above. Thus,

$$\sigma^2_r = \frac{\Sigma(X - \mu_r)^2}{N} = \frac{[(16 - 29.8)^2 + (41 - 29.8)^2 + \ldots + (39 - 29.8)^2]}{10}$$

$$= \frac{[(-13.8)^2 + (11.2)^2 + \ldots + (9.2)^2]}{10} = \frac{[190.44 + 125.44 + \ldots + 84.64]}{10}$$

$$= \frac{1,073.60}{10} = 107.36$$

The *Standard Deviation* is a measure of variation that is closely related to the variance. Specifically, for a discrete random variable X, the standard deviation is $\sigma_j = \sqrt{\sigma^2_j}$, where j = un, gr, or r.

Example #7:

$$\sigma_{un} = \sqrt{107.36} = 10.36$$

$$\sigma_{gr} = \sqrt{72.09} = 8.49$$

$$\sigma_r = \sqrt{107.36} = 10.36$$

Sometimes the ratio of the standard deviation to the arithmetic mean is used as an indicator of relative dispersion. This ratio is called the *Coefficient of Variation* (CV) and it may be used to compare the relative dispersion of two or more probability distributions.

$$CV_j = \left(\frac{\sigma_j}{\mu_j} \right), \text{ where } j = \text{un, gr, or r.}$$

Example #8:
If a group of people have a mean height of 66 inches with a standard deviation of 4 inches, but a mean weight of 150 pounds with a standard deviation of 30 pounds, one can say that heights are less varied than weights because for heights, $CV_H = 4"/66" = 0.06$, but for weights, $CV_W = 30lb/150lb = 0.20$.

A *Percentile* is a measure of central tendency or, more specifically, relative position, describing the relationship of a single value to the entire data set. δ_k is called the kth percentile and is the value such that approximately k% of the measurements are less, and (100-k)% are more. Popular percentiles are the *median* (middle value), *1st quartile* (middle value of the values to the left of median), and *3rd quartile* (middle value of the values to the right of median).

Example #9:
Consider the data for X_1:

16, 41, 25, 21, 30, 17, 29, 50, 30, 39

Rearrange the data from lowest to highest:

16, 17, 21, 25, 29, 30, 30, 39, 41, 50

Because we have an even number of numbers (10 all together) the median is the average of the two middle numbers. In this example, the two middle numbers are 29 and 30 and their average or median is: $\delta_{50} = 29.5$. (When we have an odd number of numbers the median is the middle number.) The median is also called *2nd quartile* or 50th percentile.

50% of the numbers to the left of the median (namely the numbers 16,17,21,25, and 29) have their own middle number which is called 1st quartile or 25th percentile. This number is: $\delta_{25} = 21$. Similarly, 50% of the numbers to the right of the median (namely the numbers 30, 30, 39, 41, and 50) have their own middle number which is called 3rd quartile or 75th percentile. This number is: $\delta_{75} = 39$.

Unlike the above measures, the third measure of central tendency, the *Mode*, does not always exist. If the set of data has a value that occurs more often than any of the others, then the value is said to be the mode. If two values occur the same number of times and more often than the others, the data set is said to be *bimodal* and it is denoted by λ_2. The data set is *multimodal* (denoted by λ_m, m = 3,4,5, . . .) if there are more than two values that occur with the same greatest frequency. (Unlike the mean and the median, the mode is applicable to qualitative as well as quantitative data.)

Example #10:
Consider the data for X_1:

16, 41, 25, 21, 30, 17, 29, 50, 30, 39

As it may be seen in this the data set, the value of 30 occurs twice, more often than any of the others. Thus the mode is 30, or $\lambda = 30$.

The shape of a frequency distribution can be described by its symmetry or lack of it which is called "skewness."

Skewness is zero, positive, or negative and it is measured by:

$$\pi_j = \frac{(\mu_j - \lambda)}{\sigma_j}, \text{ where } j = un, gr, or r.$$

A histogram (or probability distribution) is symmetric when $\mu_j - \lambda = 0$, positively skewed when $\mu_j > \lambda$, and negatively skewed when $\mu_j < \lambda$.

Example #11:
Consider the data for X_1 and the values for μ_j, σ_j and λ, computed above:

16, 41, 25, 21, 30, 17, 29, 50, 30, 39

$\mu_{un} = 29.8$, $\mu_{gr} = 29.9$, $\mu_r = 29.8$,

$\sigma_{un} = 10.36$, $\sigma_{gr} = 8.49$, $\sigma_r = 10.36$,

and $\lambda = 30$

$$\pi_{un} = \frac{(\mu_{un} - \lambda)}{\sigma_{un}} = \frac{(29.8 - 30)}{10.36} = -0.019$$

$$\pi_{gr} = \frac{(\mu_{gr} - \lambda)}{\sigma_{gr}} = \frac{(29.9 - 30)}{8.49} = -0.012$$

$$\pi_r = \frac{(\mu_r - \lambda)}{\sigma_r} = \frac{(29.8 - 30)}{10.36} = -0.019$$

which indicate that the X_1 distribution is negatively skewed.

Important Note: Whenever the underlying distribution is skewed, the median, instead of the mean, is a better measure of central tendency.

The shape of a frequency distribution can also be described by its "peakness" which is called *kurtosis*. Kurtosis is measured by:

$$\theta_{un} = \frac{\left[\dfrac{\Sigma (X - \mu_{un})^4}{N} \right]}{\sigma_{un}^4}$$

$$\theta_{gr} = \frac{\left[\dfrac{\Sigma f \cdot (m - \mu_{gr})^4}{N} \right]}{\sigma_{gr}^4}$$

$$\theta_r = \frac{\left[\dfrac{\Sigma (X - \mu_r)^4}{N} \right]}{\sigma_r^4}$$

When $\theta_j = 3$ the histogram is *mesokurtic*, when $\theta_j > 3$ the histogram is *lepto-kurtic*, and when $\theta_j < 3$ the histogram is *platykurtic*. (See Figure 2.3)

Figure 2.3

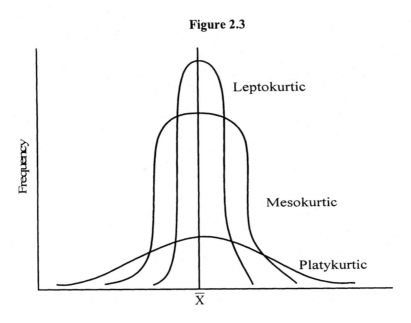

Example #12:
Consider the data for X_1: 16, 41, 25, 21, 30, 17, 29, 50, 30, 39, the values for μ_j and σ_j computed above, and Table 2.2 (for the m and f values.)

$$\theta_{un} = \frac{\left[\dfrac{\sum (X - \mu_{un})^4}{N} \right]}{\sigma_{un}^4} = \frac{\left[\dfrac{259,034.91}{10} \right]}{10.36^4} = 2.25$$

$$\theta_{gr} = \frac{\left[\dfrac{\sum f \cdot (m - \mu_{gr})^4}{N} \right]}{\sigma_{gr}^4} = \frac{\left[\dfrac{122,933.45}{10} \right]}{8.49^4} = 2.37$$

$$\theta_r = \dfrac{\left[\dfrac{\sum (X - \mu_r)^4}{N}\right]}{\sigma_r^4} = \dfrac{\left[\dfrac{259,034.91}{10}\right]}{10.36^4} = 2.25$$

Thus, the X_1 distribution is platykurtic.

2.3.1   Box-and-Whiskers Plots and Stem-and-Leaf Displays

A *Box-and-Whiskers Plot* is a pictorial display of a variable's 5-number summary. The 5-number summary of a variable consists of the values min, δ_{25}, δ_{50}, δ_{75}, and max, where min = smallest value, δ_{25} = 1st quartile, δ_{50} = median, δ_{75} = 3rd quartile, and max = largest value. The Box-and-Whiskers Plot provides the investigator with a visual impression of how a variable's values are spread out from their median.

Construction of a Box-and-Whiskers Plot requires that a box is drawn such that:

(a) the ends of the box are at 1st quartile and 3rd quartile, called the left and right hinges of the box

(b) a vertical line is drawn within the box to indicate median; and

(c) a line is drawn from the left end of the box to the min value, and another from the right end of the box to the max value (these lines are called whiskers).

Half of the X values fall in the box between its upper and lower "hinges," while the remaining half are represented by the two whiskers. The summary information provided by a box-and-whiskers plot is particularly useful when compared with similar information gathered at another place or time.

Example #13:
Consider again the data for X_1:

16, 41, 25, 21, 30, 17, 29, 50, 30, 39

The 5-number summary of X_1 is:

min = 16, δ_{25} = 21, δ_{50} = 29.5, δ_{75} = 39, and max = 50.

Figure 2.4 summarizes all this information in a Box-and-Whiskers Plot for X_1.

Figure 2.4

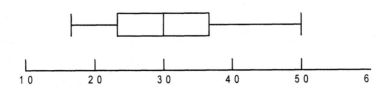

A *Stem-and-Leaf Display* is similar to a grouped histogram. Each numerical value is split into two parts: the "stem" and the "leaf" with the stem always leading. The first step in making such a display is deciding how to split each numerical value into a stem and leaf. Often the leaf will be just the last digit, or the 10s, or the 100s digit and so forth.

Example #14:
Consider once more the data for X_1:

16, 41, 25, 21, 30, 17, 29, 50, 30, 39

The Stem-and-Leaf Display, with the last digit as the leaf, is shown in Figure 2.5.

Figure 2.5

1	67
2	159
3	009
4	1
5	0

It is important to remember that the stems must follow a certain convenient arithmetic progression, up until the largest numerical value is counted. Additionally, all possible stems between the smallest and largest, even if some of them do not have values, should be included. For example, if the numbers are 1,015, 1,017, 1,023, ,2,096, 2,052, and 5,000 and the leaf is the 100s digit, then the stems are 1, 2, 3, 4, and 5 with the corresponding leafs of (0,0,0 for 1), (0,0 for 2), (nothing for 3), (nothing for 4), and (0 for 5). For the same numbers if the leaf is the 10s digit, then the stems are 10, 20, 30, 40, and 50 with the corresponding leafs of (1,1,2 for 10), (5,9 for 20), (nothing for 30), (nothing for 40), and (0 for 50).

2.4   Random Variables

2.4.1   Expected Value

If X consists of N values then the mean of X may be expressed as follows:

$$\mu = E(X) = \sum_{i=1}^{N} P_i X_i$$

where E is the expectations operator, $E(X)$ is the expected value of X, P_i is the probability that X_i occurs, and $\Sigma P_i = 1$.

Example #15:
If X consists of the numerical values 4, 5, 4, 3, 4, then

$$E(X) = 4(3/5) + 5(1/5) + 3(1/5) = 4.$$

Rule 1: If c = constant then $E(c) = c$

Example #16:
If X consists of the numerical value 7, then

$$E(X) = 7(1/1) = 7.$$

Rule 2: $E(X + c) = E(X) + c$

Example #17:
If $Y = X + 6$ and X consists of the numerical values 4, 5, 4, 3, 4, then

$$E(Y) = E(X+ 6) = E(X) + 6 = 4 + 6 = 10.$$

Rule 3: $E(cX) = cE(X)$

Example #18:
If $Y = 8X$ and X consists of the numerical values 4, 5, 4, 3, 4, then

$$E(Y) = 8E(X) = 32.$$

Rule 4: If X and Y are random variables, then $E(X + Y) = E(X) + E(Y)$

Example #19:

Example #19:
If X consists of the numerical values 4, 5, 4, 3, 4 and Y consists of the numerical values 32, 38, 31, 36, 13, then

$$E(X + Y) = E(X) + E(Y) = 4 + 30 = 34.$$

Example #20:
If $Z = 2X$ and $K = 3Y$, where X and Y are distributed as in Example #19, then

$$E(Z + K) = 2E(X) + 3E(Y) = 2(4) + 3(30) = 98.$$

Rule 5: If X and Y are independent variables, then $E(XY) = E(X)E(Y)$

Example #21:
Variables X and Y consist of the following numerical values:

X: 5, 3, 1, 3, 1, 3, 5, 1, 5 and
Y: 4, 3, 1, 12, 9, 3, 20, 15, 5.

Therefore,

$$E(XY) = E(X)E(Y) \rightarrow 8 = (3)(2.666) \rightarrow 8 = 8.$$

2.4.2   Variance

If X consists of N values then the variance of X may be expressed as follows:

$$\sigma_X^2 = VAR(X) = \sum_{i=1}^{N} P_i[X_i - E(X)]^2 \tag{1}$$

Since *variance is the average of the squared deviations from the mean*, alternatively, variance may be expressed as follows:

$$VAR(X) = E[X - E(X)]^2$$

or,

$$VAR(X) = E(X^2 + \mu^2 - 2X\mu) = E(X^2) + \mu^2 - 2E(X)\mu =$$

$$E(X^2) + \mu^2 - 2\mu^2 = E(X^2) - \mu^2 \Leftrightarrow \qquad (2)$$

$$VAR(X) = E(X^2) - [E(X)]^2$$

Example #22:
If X consists of the numerical values 4, 5, 4, 3, 4, then

$E(X) = 4(3/5) + 5(1/5) + 3(1/5) = 4.$

Following (1),
$VAR(X) = (3/5)(4\text{-}4)^2 + (1/5)(5\text{-}4)^2 + (1/5)(3\text{-}4)^2 = 2/5 = 0.4.$

Following (2),
$VAR(X) = 16.4 - 16 = 0.4.$

Rule 1: If c = constant then $VAR(c) = E[c - E(c)]^2 = E(c - c)^2 = 0$

Example #23:
If c=3, then $VAR(3) = E[3\text{-}E(3)]^2 = E(3\text{-}3)^2 = 0.$

Rule 2: $VAR(X + c) = VAR(X) + VAR(c) = VAR(X) + 0 = VAR(X)$

Example #24:
If Y = X + 6 and X consists of the numerical values 4, 5, 4, 3, 4 then,

$VAR(Y) = VAR(X) = 0.4.$

Rule 3: $VAR(cX) = c^2VAR(X)$

Proof:

$VAR(cX) =$

$E[cX - E(cX)]^2 =$

$E[cX - cE(X)]^2 =$

$$E(c^2X^2 + c^2\mu^2 - 2c^2X\,\mu) =$$

$$c^2E(X^2) + c^2\mu^2 - 2c^2\mu^2 =$$

$$c^2E(X^2) - c^2\mu^2 =$$

$$c^2[E(X^2) - \mu^2] = c^2\{E(X^2) - [E(X)]^2\} =$$

$$c^2VAR(X).$$

Example #25:
If Y = 8X and X consists of the numerical values 4, 5, 4, 3, 4, then

$$VAR(Y) = 8^2VAR(X) = 64(0.4) = 25.6.$$

Example #26:
If Y = 12 + 2X and E(X) = 10 then

$$VAR(Y) = 2^2VAR(X) = 40.$$

Example #27:
If $Y = 12 + 2X^2$ and X consists of the numerical values 4, 5, 4, 3, 4, then

$$VAR(Y) = 4VAR(X^2) = 4(25.84) = 103.36.$$

Example #28:
If E(X) = 2, VAR(X) = 6 and $Y = 3 + 9X^2$ show that E(Y) = 93.

$E(Y) = 3 + 9E(X^2)$. From (2), $E(X^2) = VAR(X) + [E(X)]^2 = 6 + 2^2 = 10$.

Therefore, E(Y) = 3 + 9(10) = 93.

Rule 4: The variance of the sum of two random variables X and Y is

$$VAR(X + Y) = VAR(X) + VAR(Y) + 2COV(X,Y)$$

Note: COV(X,Y) = Covariance between X and Y;
       COV(X,Y) = E(XY) − E(X)E(Y), or COV(X,Y) = E[X − E(X)][Y − E(Y)].

As the last formula indicates, covariance is the expectation of the product of variables X and Y when each is measured as deviation around its mean. If variables X and Y are independent, COV(X,Y) = 0 and VAR(X + Y) = VAR(X) + VAR(Y). More on Covariance follows below.

Proof:

$$VAR(X + Y) =$$

$$E[(X + Y) - E(X + Y)]^2 =$$

$$E[X + Y - E(X) - E(Y)]^2 =$$

$$E\{[X - E(X)] + [Y - E(Y)]\}^2 =$$

$$E\{[X - E(X)]^2 + [Y - E(Y)]^2 + 2[X - E(X)][Y - E(Y)]\} =$$

$$E[(X - E(X)]^2 + E[(Y - E(Y)]^2 + 2E[(X - E(X)][(Y - E(Y)] =$$

$$VAR(X) + VAR(Y) + 2COV(X, Y).$$

Example #29:

Consider random variables X and Y. If X consists of the numerical values 4, 5, 4, 3, 4 and Y consists of the numerical values 32, 38, 31, 36, 13 then VAR (X+ Y) may be computed with the assistance of the table below as follows:

X	Y	XY
4	32	128
5	38	190
4	31	124
3	36	108
4	13	52
E(X) = 4	E(Y) = 30	E(XY) = 120.4
VAR(X) = 0.4	VAR(Y) = 78.8	COV(X,Y) = 0.4

Therefore, VAR(X+Y) = 0.4 + 78.8 + 2(0.4) = 80

Rule 5: If $\overline{X}$ is the sample mean of a random variable with mean μ and variance

σ_X^2, then $\text{VAR}(\overline{X}) = \dfrac{\sigma_X^2}{N}$;

Proof:

$$\text{VAR}(\overline{X}) = \text{VAR}\left(\frac{1}{N}\sum_{i=1}^{N}X_i\right) =$$

$$\left(\frac{1}{N}\right)^2 \text{VAR}\left(\sum_{i=1}^{N}X_i\right) =$$

$$\left(\frac{1}{N}\right)^2 \sum_{i=1}^{N}\text{VAR}(X_i) =$$

$$\left(\frac{1}{N}\right)^2 N\,\text{VAR}(X_i) =$$

$$\frac{\text{VAR}(X_i)}{N} = \frac{\sigma_X^2}{N}.$$

Example #30:

Consider population X which consists of the following three (N=3) numerical values: 2, 5, 3. The mean and variance of X are $\mu = 3.33$ and $\sigma^2 = 1.55$. Sampling with replacement, draw all possible samples of size n=3 (27 in total) and compute their averages (for a start, see the table below); in turn, compute the variance of all averages. The variance of all averages (computed with the population formula) should be equal to 0.52.

	Sample	Sample Average
1	2,2,2	2
2	2,2,5	3
3	2,2,3	2.33
.	.	.
.	.	.
.	.	.
27	...	...

Notice also that the average of all averages is 3.33, identical to the mean of the population.

Note: As we shall learn in next Chapter, the set of sample averages is called *Sampling Distribution of the Mean* (SDM). The SDM's average is equal to the mean of the population, it is normally distributed and, for a finite population, its variance is equal to the ratio of the variance of the population to its size.

Rule 6: The average of all (or a relatively large number of) sample variances is equal to (or approximately equal to) the variance of the population.

Example #31:
Consider the population in the previous Example. Sampling with replacement, draw all possible samples of size n=3 (27 in total) and compute their variances (see the table below); in turn, compute the average of all variances. The average of all variances should be equal to the variance of the population or, 1.55.

	Sample	Sample Variance
1	2,2,2	0
2	2,2,5	2
3	2,2,3	0.22
.	.	.
.	.	.
.	.	.
27	...	...

The set of sample variances is called *Sampling Distribution of the Variance* (SDV) and it is exponentially distributed.

2.4.3   Covariance

In general, covariance is a measure of the linear relationship between two random variables and its value depends on the units in which variables are measured. In addition to the formulas above, an alternative way to express covariance is as follows:

$$COV(X, Y) = \sum_{i=1}^{N} [X_i - E(X)][Y_i - E(Y)] = \sigma_{XY},$$

where σ_{xy} = *Population Covariance*. Alternatively,

$$\sigma_{XY} = \frac{\sum_{i=1}^{N}[X_i - E(X)][Y_i - E(Y)]}{N}$$

Note: Sample Covariance is

$$S_{XY} = \frac{\sum_{i=1}^{n}[X_i - E(X)][Y_i - E(Y)]}{n-1}, \text{ where } n = \text{sample size.}$$

Figures 2.6 and 2.7 are scatter-plots graphically illustrating positive and negative covariance. When both variables are always observed to be above and below their respective means, the covariance is positive; if one is below its mean when the other is above its mean (and vice-versa), the covariance is negative.

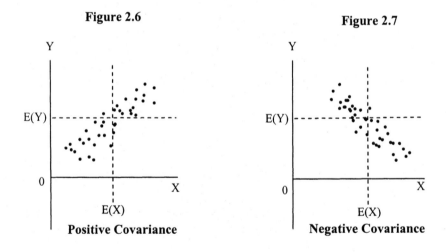

Figure 2.6

Positive Covariance

Figure 2.7

Negative Covariance

Rule 1: If c and d are constants, COV(c, d) = 0

Example #32:
If c = 3 and d = 6, then

$$COV(3, 6) = E\{[3 - E(3)][6 - E(6)]\} = E[(3 - 3)(6 - 6)] = 0.$$

Rule 2: When two random variables X and Y are independent,

COV(X,Y) = 0 and their scatter-plot looks as in Figure 2.8

Figure 2.8

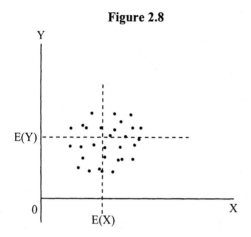

Proof:

COV(X, Y) =

$E[X - E(X)][Y - E(Y)] =$

$E[XY - XE(Y) - YE(X) + E(X)E(Y)] =$

$E(XY) - E(X)E(Y) - E(Y)E(X) + E(X)E(Y) =$

$E(XY) - E(X)E(Y);$

since, by Rule 5 of Expected Value, E(XY) = E(X)E(Y),

COV(X, Y) = 0.

Rule 3: COV(X, Y) = COV(Y, X)

Rule 4: If c is a constant, COV(X, c) = 0

Rule 5: If c and d are constants, COV(X + c, Y + d) = COV(X, Y)

Example #33:
If K = 10 + X and L = 2 + Y, then

$$COV(K, L) = COV(10 + X, 2 + Y) = COV(X, Y)$$

Example #34:
Compute COV(K, L) when K = 10 + X, L = 2 + Y and X consists of the numerical values 4, 5, 4, 3, 4 while Y consists of the numerical values 32, 38, 31, 36, 13.

From the above example,

$$COV(K, L) = COV(X, Y) = E(XY) - E(X)E(Y) = 120.4 - (4)(30) = 0.4.$$

Rule 6: If c and d are constants, COV(cX, dY) = cdCOV(X, Y)

Example #35:
If K = 5X and L = 8Y, then

$$COV(K, L) = COV(5X, 8Y) = 40COV(X, Y)$$

Example #36:
If K = 2 + 3X and L = 6 + 5Y, then

$$COV(K, L) = COV(2 + 3X, 6 + 5Y) = COV(3X, 5Y) = 15COV(X, Y).$$

Rule 7: COV(X + Y, Z) = COV(X, Z) + COV(Y, Z)

Example #37:
If K = 5X, L = 2Y and M = 4Z, then

$$COV(K + L, M) = COV(5X + 2Y, 4Z) = 20COV(X, Z) + 8COV(Y, Z).$$

Rule 8: COV(X, X) = VAR(X)

Proof:

$$COV(X, X) = E[X - E(X)][X - E(X)] = E[X - E(X)]^2 = VAR(X)$$

2.4.4   Correlation

Like covariance, correlation measures the strength of the relationship between two variables. Unlike covariance, correlation is normalized to range between "-1" and "+1" and, hence, it is "scale-free". It is computed as the ratio of covariance between two variables to the product of their standard deviations; formally,

$$\rho_{xy} = CORR(X, Y) = \frac{COV(X, Y)}{\sigma_X \sigma_Y},$$

where ρ_{xy} = *Population Correlation.*

Note: Sample Correlation is

$$r_{xy} = \frac{S_{xy}}{S_x S_y}, \text{ where}$$

S_{xy} = sample covariance,
S_x = sample standard deviation of X, and
S_y = sample standard deviation of Y.

Correlation is positive when $0 < \rho_{xy} \leq +1$, negative when $-1 \leq \rho_{xy} < 0$, and zero when $\rho_{xy} = 0$. The closer to the absolute value of one (zero), the stronger (weaker) the correlation is between the two variables.

Consider the following correlation examples: generally speaking, the correlation between Grade Point Average (GPA) and the Student Achievement Test (SAT) should be positive; the correlation between the inflation rate based on the Gross Domestic Product (GDP) Implicit Price Deflator and the inflation rate based on the Consumer Price Index (CPI) is expected to be positive as well. It has been empirically observed that the correlation between sales of movie tickets and the economy's business cycle is negative; in slow-downs and recessions, sales of tickets rise; in periods of prosperity and growth, sales of tickets drop.

Like with covariance, one may use scatter plots as in Figure 2.9 below to graphically illustrate correlation. In general, the denser the scatter plot the stronger the correlation.

Rule 1: If c and d are constants, then CORR(X + c, Y + d) = CORR(X, Y)

Proof:

$$CORR(X + c, Y + d) =$$

$$\frac{COV(X+c, Y+d)}{[VAR(X + c)]^{0.5}[VAR(Y + d)]^{0.5}} =$$

$$\frac{COV(X, Y)}{[VAR(X)]^{0.5}[VAR(Y)]^{0.5}} =$$

$$CORR(X, Y).$$

Figure 2.9

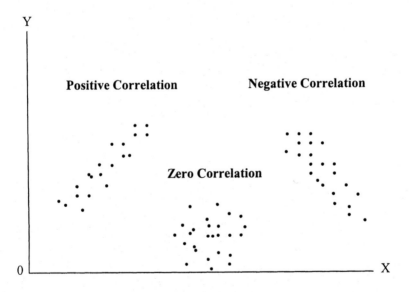

Rule 2: When $K = a + bX$ and $L = c + dY$, where a, b, c and d are constants,

$$CORR(a + bX, c + dY) = CORR(X, Y)$$

Proof:

$$CORR(K, L) = CORR(a + bX, c + dY) =$$

$$\frac{COV(a + bX, c + dY)}{[VAR(a + bX)]^{0.5} [VAR(c + dY)]^{0.5}} =$$

$$\frac{bdCOV(X, Y)}{[VAR(X)]^{0.5} [VAR(Y)]^{0.5}} =$$

$$\frac{COV(X, Y)}{[VAR(X)]^{0.5} [VAR(Y)]^{0.5}} =$$

CORR(X, Y).

2.5 GENERAL QUESTIONS AND ANSWERS

Question 1. Construct the probability distribution associated with the rolling of a symmetric triangular pyramid. (The triangular pyramid has four sides.)

a. Is the distribution "classical," "relative frequency," or "subjective?" Why or why not?
b. Draw the histogram.
c. Assume that the four possibilities are associated with the following $ amounts: 10, 12, 15, and 22. Compute the mean and standard deviation.

Answers for #1:
The distribution is:

Possibilities:	SIDE1	SIDE2	SIDE3	SIDE4
Probabilities:	1/4	1/4	1/4	1/4

a. The distribution is "classical" because the possible outcomes are mutually exclusive and equally likely.

b. The histogram is:

Frequency over 4

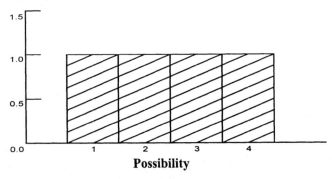

Possibility

c. Mean = 14.75, Standard Deviation = 4.55

Question 2. Construct the probability distribution X associated with the rolling of two symmetric triangular pyramids i and j.

Let X = i+j.

a. Is the distribution "classical," "relative frequency," or "subjective?" Why or why not?

b. Draw the histogram.

c. Let's play "Chuck-A-Luck." I am the "house" and you are the "player." To play this game you have to pay me a $1. Choose your favorite number between 1 and 4. Roll the pyramids. If your favorite number appears twice, I will return to you $3, and if it appears once, I'll return to you $2. Construct the probability distribution associated with this game, and draw the histogram. Compute the mean and standard deviation. Would it be a good idea to play this game? Why or why not? Is the fee of $1 fair? What fee would make this game fair?

d. Another "house" offers to you the same game at a fee of $0.8 and returns of $3.50 and $1.50 respectively. Which one of the two games will you play? Why?

Answers for #2:
Sample Space:

```
1,1  2,1  3,1  4,1
1,2  2,2  3,2  4,2
1,3  2,3  3,3  4,3
1,4  2,4  3,4  4,4
```

The distribution is: $X(i,j) = i+j$ or,

```
2   3   4   5
3   4   5   6
4   5   6   7
5   6   7   8
```

Possibilities:	Probabilities:
2	1/16
3	2/16
4	3/16
5	4/16
6	3/16
7	2/16
8	1/16

a. The distribution for X is of the "relative frequency" type because although the possible outcomes are mutually exclusive not all are equally likely.

b. The histogram is:

Frequency over 16

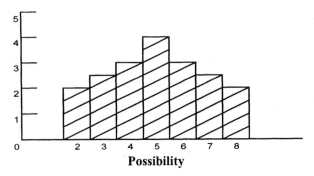

Possibility

c. The probability distribution associated with this game is:

Possibilities (or payoffs)	Probabilities
$2	1/16
$1	6/16
-$1	9/16

The histogram is:

Frequency over 16

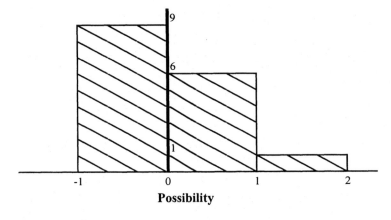

Possibility

Mean = -0.063, Standard Deviation = 1.03

This is not a fair game (it is more than fair to the house), and since the standard deviation is greater than zero it is also risky. A risk-averse individual may not want to play this game.

The fee (F) that makes the game fair is the fee that corresponds to $\mu = 0$. Since,

$$\mu = (3 - F)(1/16) + (2 - F)(6/16) + (0 - F)(9/16)$$

then, to get the fair fee, just set the right side of the above equation equal to zero and solve for F. Or,

$$(3 - F)(1/16) + (2 - F)(6/16) + (0 - F)(9/16) = 0$$

$$\Rightarrow F = 0.9375.$$

d. Mean = -0.019, Standard Deviation = 0.98. This game is more fair and less risky than the other.

Question 3. In tennis there are four possible outcomes of your first serve: A = ace, B = service winner, C = good serve but not ace (neither service winner), and D = miss. In 100 trials you had 3 aces, 22 service winners, and 50 misses. Summarize the results. What is your first serve percentage? Is it low or high?

Answers for #3:

Possibilities:	Probabilities:
A	3/100
B	22/100
C	25/100
D	50/100

My first serve percentage = (3+22+25)/100 = 0.5 or 50% which is neither high nor low.

Question 4. Sixty years ago, 4 million out of 10 million high school graduates were not able to attend college. Today the population of high school graduates is 14 million and it is estimated that 9 million of them will attend college. What are the probabilities of attending and not attending college sixty years ago and today? Has the probability of attending college increased over the years?

Answers for #4:
Let:

A = attend college, B = unable to attend college 60 years ago:

Possibilities	Probabilities
A	6/10 = 0.6
B	4/10 = 0.4

Today:

Possibilities	Probabilities
A	9/14 = 0.64
B	5/14 = 0.36

As it may be seen A has increased from 0.6 to 0.64.

Question 5. The probability of the New England Patriots making it to the playoffs this year is 0.45. What are the odds in favor of the team making it? What are the odds against the team making it?

Answers for #5:

Odds in favor = $0.45/(1-0.45) = 0.82$

Odds against = $0.55/(1-0.55) = 1.22$

Question 6. The odds of the Greyhounds making it to the elite 8 this year are 3 to 4. What are the probabilities of making it and not making it?

Answers for #6:
Let:
P_A = probability of making it,
P_B = probability of not making it, and
O = odds.

$O_A = P_A/P_B \Rightarrow 3/4 = P_A/(1-P_A) \Rightarrow P_A = 0.43.$

Therefore, $P_B = 0.57$.

Question 7. (Do by hand).

The data for sample Y1 is:

> 55, 42, 30, 75, 34, 39, 55, 42, 29, 42,
> 70, 50, 51, 53, 55, 23, 41, 55, 66, 69,
> 67, 64, 63, 42, 33

a. Construct the histogram and compute the mean, the variance, the standard deviation, the coefficient of variation, the 25th percentile, the median, the 75th percentile, the mode, the skewness coefficient, the kurtosis coefficient, construct the Box-and-Whiskers Plot, and provide a Stem-and-Leaf Display.

b. Group the data into 8 intervals and construct the histogram.

Answers for #7:

a. The histogram is:

Frequency over 25

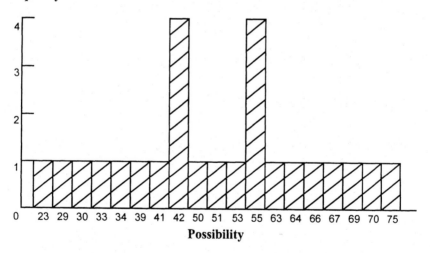

Using the formulas for sample statistics, with s divided by (n-1), we have:

Mean = 49.8
Variance = 207.417
Standard Deviation = 14.402

Coefficient of Variation = 0.289
25th Percentile = 41
Median = 51
75th Percentile = 63
Bimodal (42 and 55)
Skewness = -0.030
Kurtosis = -1.013

The Box-and-Whiskers Plot is:

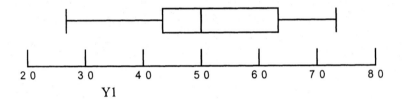

The Stem-and-Leaf Display is:

2	39
3	0349
4	H 12222
5	M 0135555
6	H 34679
7	05

b.

INTERVALS	FREQUENCY	PROBABILITY
23 – 29	2	2/25
30 – 36	3	3/25
37 – 43	6	6/25
44 – 50	1	1/25
51 – 57	6	6/25
58 – 64	2	2/25
65 – 71	4	4/25
72 – 78	1	1/25

Question 8. The local weather forecaster gives a 0.8 chance of thunderstorms. A pilot uses this forecast as a "prior" probability. The pilot also considers the degree of arthritic pain experienced. From a survey of 200 days the following information was obtained.

	Thunderstorms	No Thunderstorms	Total
Severe Arthritic Pain	80	10	90
Mild Arthritic Pain	20	90	110
Total	100	100	200

Right now, the pain is severe. What is the pilot's posterior probability of thunderstorms.

Answer for #8:

The sensitivity measure indicates severity of arthritic pain for days when there are thunderstorms. Here, sensitivity = 80/100 = 0.8.

The specificity measure indicates the mildness of arthritic pain for days when there are no thunderstorms. Here, specificity = 90/100 = 0.9.

Lastly, "prior" odds = 0.8/0.2 = 4. Thus, applying the "Bayes Equation"

$$\text{Posterior Odds} = 4 \, [\, 0.8 \, / \, (1\text{-}0.9) \,] = 32$$

From this,

$$\text{"Posterior" probability of thunderstorms} = 32 \, / \, (1\text{+}32) = 0.97$$

2.6   HEALTH AND MEDICAL QUESTIONS AND ANSWERS

Question 1. 500 people used a home test for HIV. Results of the home test are given in the matrix below

	HIV Present	HIV Absent	Total
Positive Test	35	25	60
Negative Test	5	435	440
Total	40	460	500

a.   What is the sensitivity of the test?

b.   What is the specificity of the test?

c.   What is the positive predictive value?

d.   What is the negative predictive value?

Answers for #1:

a.   35/40 = 0.875

b.   435/460 = 0.946

c.   35/60 = 0.853

d.   435/440 = 0.989

Question 2. A drug company is developing a new pregnancy-test kit for use on an outpatient basis. The company uses the test on 200 women. Test results are given in the following matrix

	Pregnant	Not Pregnant	Total
Positive Test	95	1	96
Negative Test	5	99	104
Total	100	100	200

a.   What is the sensitivity of the test?

b.   What is the specificity of the test?

c. What is the positive predictive value of the test?

d. Suppose the "cost" of a false negative (2C) is twice that of a false positive (C) (Since for a false negative prenatal care would be delayed during the first trimester of pregnancy). If the standard home pregnancy test kit (made by another drug company) has a sensitivity of 0.89 and a specificity of 0.98, then which test (new or standard) is expected to cost less per woman using it and by how much?

Answer for #2:

a. $95/100 = 0.95$

b. $99/100 = 0.99$

c. $95/96 = 0.989$

d. New test:

P of false negative = 1-sensitivity = 0.05
P of false positive = 1-specificity = 0.01

$E(Cost) = (0.05)(2C)+(0.01)(C) = 0.11C$

Standard test:

P of false negative = 1-sensitivity = 0.11
P of false positive = 1-specificity = 0.02

$E(Cost) = (0.11)(2C)+(0.02)(C) = 0.24C$

Since $0.11C < 0.24C \Rightarrow$ new test costs less.

Question 3. A test has a sensitivity of 0.90 and a specificity of 0.92. The prior odds are 0.05. What is the "Bayesian/Posterior" probability?

Answer for #3:

Posterior Odds = $0.05[0.90/(1-0.92)] = 0.5625$
Posterior Probability = $0.5625/(1-0.5625) = 0.36$

Question 4. Assume another patient has all the symptoms of porphyria. Based on educated clinical judgement, you feel that the probability that the patient has the disease is 0.3. Moreover, a population screening test of 1,000 individuals yields the following test results

	Disease Present	Disease Absent	Total
Positive Test	246	26	272
Negative Test	54	674	728
Total	300	700	1,000

Calculate the "Bayesian/Posterior" odds and probability.

Answers for #4:

Sensitivity = 246/300 = 0.820
Specificity = 674/700 = 0.963

Prior Odds = 0.3/(1-0.3) = 0.429

Bayesian/Posterior Odds = 0.429[0.82/(1-0.963)] = 9.51

Bayesian/Posterior Probability = 9.51/(1+9.51) = 0.90

2.7   EXPECTED VALUE, VARIANCE, COVARIANCE, AND CORRELATION QUESTIONS AND ANSWERS

Question 1. Let $Y = 5X + 2$, $E(X) = 18$, $VAR(X) = 4$.
Compute $E(Y)$ and $VAR(Y)$.

Answers for #1:

$E(Y) = 5E(X) + 2 = 90 + 2 = 92$

$VAR(Y) = 5^2 VAR(X) + 0 = 100$.

Question 2. Let $Y = -3 + 2X - 5Z$, $E(X) = 4$, $VAR(X) = 2$ and [Z: 2, 6, 1].
Compute $E(Y)$ and $VAR(Y)$.

Answers for #2:

$E(Y) = -3 + 2E(X) - 5E(Z) = -3 + 8 - 5(3) = -10$

$VAR(Y) = 4VAR(X) - 25VAR(Z) = 8 - 25(4.67) = -108.67$.

Question 3. Let $Y = 5X^2 + 2$ and [X: 8, 1, 5, 6].
Compute $E(Y)$ and $VAR(Y)$.

Answers for #3:

$E(Y) = 5E(X^2) + 2 = 5(31.5) + 2 = 159.20$

$VAR(Y) = 25VAR(X^2) = 25(512.25) = 12,806.25$.

Question 4. Let $Y = 4 + 10X^2$, $E(X) = 3$, $VAR(X) = 7$ and $VAR(X^2) = 42$.
Compute $E(Y)$ and $VAR(Y)$.

Answers for #4:

From $VAR(X) = E(X^2) - [E(X)]^2 \rightarrow E(X^2) = VAR(X) + [E(X)]^2 = 7 + 9 = 16$

Hence, $E(Y) = 4 + 10E(X^2) = 4 + 10(16) = 164$
and $VAR(Y) = 100VAR(X^2) = 4,200$.

Question 5. Consider random variables X and Y. If X consists of the numerical values 3, 7, 2, 5, 8 and Y consists of the numerical values 5, 2, 4, 2, 1. Compute VAR (X+ Y).

Answers for #5:

X	Y	XY
3	5	15
7	2	14
2	4	8
5	2	10
8	1	8
E(X) = 5	E(Y) = 2.8	E(XY) = 11
VAR(X) = 5.2	VAR(Y) = 2.16	COV(X,Y) = -0.232

Therefore, VAR(X+Y) = 5.2 + 2.16 + 2(-0.232) = 6.896.

Question 6. Let Y = 4X + 3, Z = 6K + 7, E(X) = 3, E(K) = 4, COV(X, K) = 5, VAR(X) = 6, VAR(K) = 8, (assuming all variables are random).

Compute: E(Y), E(Z), COV(Y, Z), CORR(Y, Z) and VAR(Y + Z).

Answers for #6:

E(Y) = 4(3) + 3 = 15

E(Z) = 6(4) + 7 = 31

COV(Y, Z) = COV(4X + 3, 6K + 7) = (4)(6)COV(X, K) = 24(5) = 120

CORR(Y, Z) = CORR(4X + 3, 6K + 7) =

$$\frac{COV(X, K)}{[VAR(X)]^{0.5}\,[VAR(Y)]^{0.5}} =$$

$$\frac{5}{(2.45)(2.83)} = 0.721$$

$VAR(Y + Z) = VAR(Y) + VAR(Z) + 2COV(Y, Z) =$

$16VAR(X) + 36VAR(K) + 2(120) = 16(6) + 36(8) + 240 = 624.$

Question 7. Let $Y = X + 1$, $Z = 2K + 2$, $E(X) = 5$, $E(K) = 4$, $COV(X, K) = 6$, $VAR(X) = 4$, $VAR(K) = 5$, (assuming all variables are random).
Compute: $E(Y)$, $E(Z)$, $COV(Y, Z)$, $CORR(Y, Z)$ and $VAR(Y + Z)$.

Answers for #7:

$E(Y) = 5 + 1 = 6$

$E(Z) = 2E(K) + 2 = 10$

$COV(Y, Z) = COV(X + 1, 2K + 2) = 2COV(X, K) = 2(6) = 12$

$CORR(Y, Z) = CORR(X + 1, 2K + 2) =$

$$\frac{COV(X, K)}{[VAR(X)]^{0.5} [VAR(Y)]^{0.5}} =$$

$$\frac{6}{(2.00)(2.24)} = 1.340$$

$VAR(Y + Z) = VAR(Y) + VAR(Z) + 2COV(Y, Z) =$

$VAR(X) + 4VAR(K) + 2(12) = 4 + 4(5) + 24 = 48.$

Question 8. Consider population X which consists of the following three (N=3) numerical values: 5, 3, 9. The mean and variance of X are, respectively, $\mu = 5.67$ and $\sigma^2 = 6.22$. Sampling with replacement, draw all possible samples of size n=3 and compute the following: sample averages ($\overline{X}$) and sample variances

(S_X^2); in turn, show that $E(\overline{X}) = \mu$, $VAR(\overline{X}) = \dfrac{\sigma^2}{N}$, and $\overline{S}_X^2 = \sigma^2$, where

$\overline{S}_X^2$ = average of variances.

Answers for #8:

SAMPLES			$\overline{X}$	S_X^2
5	5	5	5.000000	0.000000
5	5	3	4.333333	1.333333
5	5	9	6.333333	5.333333
5	3	5	4.333333	1.333333
5	3	3	3.666667	1.333333
5	3	9	5.666667	9.333333
5	9	5	6.333333	5.333333
5	9	3	5.666667	9.333333
5	9	9	7.666667	5.333333
3	5	5	4.333333	1.333333
3	5	3	3.666667	1.333333
3	5	9	5.666667	9.333333
3	3	5	3.666667	1.333333
3	3	3	3.000000	0.000000
3	3	9	5.000000	12.000000
3	9	5	5.666667	9.333333
3	9	3	5.000000	12.000000
3	9	9	7.000000	12.000000
9	5	5	6.333333	5.333333
9	5	3	5.666667	9.333333
9	5	9	7.666667	5.333333
9	3	5	5.666667	9.333333
9	3	3	5.000000	12.000000
9	3	9	7.000000	12.000000
9	9	5	7.666667	5.333333
9	9	3	7.000000	12.000000
9	9	9	9.000000	0.0000000

Hence,

$$E(\overline{X}) = 5.67$$

$$VAR(\overline{X}) = \frac{\sum_{i=1}^{27}[\overline{X}_i - E(\overline{X})]^2]}{27} = 2.07$$

$$\overline{S}_X^2 = 6.22.$$

Question 9. Consider *populations* [X: 7, 8, 12, 13, 14, 15, 18, 20] and [Y: 22, 19, 18, 15, 14, 13, 12, 10]. Compute the covariance and correlation between X and Y; in turn, graph the scatter plot with Y on the vertical axis.

Answers for #9:

E(X) = 13.38, E(Y) = 15.38, E(XY) = 190.50, SD(X) = 4.18, SD(Y) = 3.74;

therefore,

COV(X, Y) = E(XY) – E(X)E(Y) = 190.50 – 205.78 = - 15.28

CORR(X, Y) = - COV(X, Y) / [SD(X)SD(Y)] = 15.28 / 15.63 = - 0.98

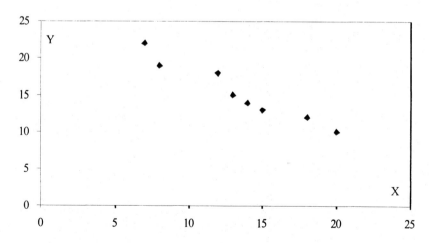

Question 10. Consider *samples*

[X: 8, 14, 16, 17, 20, 22, 15, 20] and

[Y: 9, 11, 13, 18, 19, 21, 14, 16].

Compute the covariance and correlation between X and Y; in turn, graph the scatter plot with Y on the vertical axis.

Answers for #10:

$$S_{XY} = \frac{\displaystyle\sum_{i=1}^{8}[X_i - E(X)][Y_i - E(Y)]}{n-1} = 16.50$$

$$r_{xy} = \frac{S_{xy}}{S_x S_y} = \frac{16.50}{(4.41)(4.12)} = 0.91$$

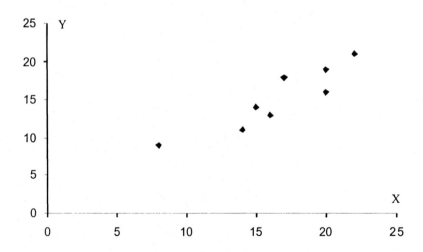

Question 11. Numerical values for samples X and Y are as follows :

[X: 2, 7, 8, 10, 4, 9, 11, 4, 8, 5]

[Y: 3, 6, 9, 12, 3, 8, 10, 5, 7, 6]

Let K = 20 + 2X and L = 10 + 1.5Y.

Show that the correlation between K and L is equal to the correlation between X and Y or that $r_{KL} = r_{XY}$.

Answers for #11:

$$r_{KL} = \frac{S_{KL}}{S_K S_L} = \frac{23.60}{(5.87)(4.38)} = 0.92$$

and

$$r_{XY} = \frac{S_{XY}}{S_X S_Y} = \frac{7.87}{(2.94)(2.92)} = 0.92$$

X	Y	K	L
2	3	24	14.50
7	6	34	19.00
8	9	36	23.50
10	12	40	28.00
4	3	28	14.50
9	8	38	22.00
11	10	42	25.00
4	5	28	17.50
8	7	36	20.50
5	6	30	19.00

Question 12. Consider the averages and standard deviations for holding period returns (HPR) of a portfolio consisting of two stocks: y and x.

(a) Based on the summary measures and formulae below, numerically derive, and graph in the same diagram, three Efficient Frontiers of the portfolio as follows: for the first efficient frontier let the correlation between the y and x HPRs be $\rho_{yx} = 1$, for the second let $\rho_{yx} = 0$, and for the third let $\rho_{yx} = -0.95$. The Efficient Frontier is a locus of points of portfolios between Portfolio Expected Return $E(P_R)$ and Portfolio Variance (σ^2_P) based on all possible spending allocations (weights: w_y, w_x) of an investor's fixed budget. (In the case of this y, x portfolio, $w_y + w_x = 1$.)

$$\mu_y = 0.10, \quad \sigma_y = 0.16$$
$$\mu_x = 0.27, \quad \sigma_x = 0.22$$

$$E(P_R) = w_y\mu_y + w_x\mu_x; \text{ or, } E(P_R) = w_y\mu_y + (1-w_y)\mu_x$$

$$\sigma^2_P = w_y\sigma_y^2 + (1-w_y)\sigma_x^2 + 2w_y(1-w_y)\sigma_y\sigma_x\rho_{yx}$$

(b) Describe some lessons from the analysis. [Hints: (1) How much would you spend on y?; how much on x?; why? (2) What is the impact of correlation? Alternatively, which one of the three frontiers would you prefer and why?]

Answers for #12:

(a) The table below displays arbitrarily chosen values for w_y and the correspon-ding values of $E(P_R)$ and σ^2_P subject to various correlation values. In turn, the three Efficient Frontiers that correspond to the data in the table are gra-phed in the figure below:

$[E(P_R) \text{ vs. } \sigma^2_P \text{ with } \rho_{yx} = -0.95],$

$[E(P_R) \text{ vs. } \sigma^2_P \text{ with } \rho_{yx} = 0], \text{ and}$

$[E(P_R) \text{ vs. } \sigma^2_P \text{ with } \rho_{yx} = +1].$

w_y	$E(P_R)$	σ^2_P $\rho_{yx}= -0.95$	σ^2_P $\rho_{yx}= 0$	σ^2_P $\rho_{yx}= +1$
1.00	*0.1000*	0.025600	0.025600	*0.025600*
0.95	0.1085	0.020048	0.023225	0.026569
0.90	0.1170	0.015201	0.021220	0.027556
0.85	0.1255	0.011058	0.019585	0.028561
0.80	0.1340	0.007619	0.018320	0.029584
0.75	0.1425	0.004885	0.017425	0.030625
0.70	0.1510	0.002855	0.016900	0.031684
0.65	*0.1595*	0.001530	*0.016745*	0.032761
0.60	*0.1680*	*0.000909*	0.016960	0.033856
0.55	0.1765	0.000992	0.017545	0.034969
0.50	0.1850	0.001780	0.018500	0.036100
0.45	0.1935	0.003272	0.019825	0.037249
0.40	0.2020	0.005469	0.021520	0.038416
0.35	0.2105	0.008370	0.023585	0.039601
0.30	0.2190	0.011975	0.026020	0.040804
0.25	0.2275	0.016285	0.028825	0.042025
0.20	0.2360	0.021299	0.032000	0.043264
0.15	0.2445	0.027018	0.035545	0.044521
0.10	0.2530	0.033441	0.039460	0.045796
0.05	0.2615	0.040568	0.043745	0.047089
0.00	*0.2700*	*0.048400*	*0.048400*	*0.048400*

(b)

(1) Consider the Efficient Frontier that corresponds to $\rho_{yx}= + 1$ (last column in the table and straight line frontier in the graph.) Obviously, the best location in terms of risk would be when $w_y = 1$ or when you 'spend' your entire investing budget on stock y; at that point $E(P_R) = 0.1$ and $\sigma^2_P = 0.0256$. *To experience a higher return you have to be able to absorb more risk* by moving northeast along the frontier; eventually, the highest return will be achieved when you 'spend' all of your investing budget on stock x and make $E(P_R) = 0.27$ but at the highest risk of $\sigma^2_P = 0.0484$.

(2) Consider the Efficient Frontier that corresponds to $\rho_{yx}= 0$ (fourth column in the table and middle frontier in the graph.) Notice that risk decreases as you allocate more funds on x and less on y while, simultaneously, portfolio return increases. Obviously, the best location in terms of risk would be when $w_y = 0.65$ or when you 'spend' 65% of your investing budget on stock y and 35% on x; at that point $E(P_R) = 0.1595$ and $\sigma^2_p = 0.016745$. Notice that if you are currently at point A on this frontier, by moving to point B you improve both in terms of risk and portfolio return. If you decide to move northeast beyond point B, you shall improve in terms of return but not improve in terms of risk (a move that you may desire to make only if you can afford more risk.) Notice also that the $\rho_{yx}= 0$ frontier outperforms the $\rho_{yx}= +1$ frontier in terms of risk across the board.

(3) Consider the Efficient Frontier that corresponds to $\rho_{yx}= - 0.95$ (third column in the table and outmost left frontier in the graph.) Notice that risk decreases as you allocate more funds on x and less on y while, simultaneously, portfolio return increases. Obviously, the best location in terms of risk would be when $w_y = 0.60$ or when you 'spend' 60% of your investing budget on stock y and 40% on x; at that point $E(P_R) = 0.1680$ and $\sigma^2_p = 0.000909$. Notice that if you are currently at a point on the negatively- sloped portion of this frontier, by moving to point C you improve both in terms of risk and portfolio return. If you decide to move northeast beyond point C, you shall improve in terms of return but not improve in terms of risk (a move that you may desire to make only if you can afford more risk.) Notice also that the $\rho_{yx}= - 0.95$ frontier outperforms the $\rho_{yx}= 0$ frontier in terms of risk across the board. Hence, the lower towards "- 1" the correlation between stocks, the further to the left the frontier moves and the better off the investor becomes; for example, an investor moving from B to C gains in both risk and return. Obviously, risk is minimized when $\rho_{yx}= - 1$.

In brief, some of the lessons in association with this example are as follows:

- Given your investing budget (the amount of money you have decided to invest in stocks y and x) you can compute the spending percentages that minimize portfolio risk.

- When $\rho_{yx}= +1$, the higher the return you desire from a portfolio of stocks y and x the higher the risk you have to absorb.

- When $\rho_{yx} \leq 0$ and the Efficient Frontier is negatively-sloped, you may achieve higher return at lower risk by relocating northwest on the frontier. When $\rho_{yx} \leq 0$ and the Efficient Frontier is positively sloped, you may achieve higher return at higher risk by relocating northeast on the frontier.

- The more negatively correlated the stocks are the further to the left the Efficient Frontier moves and therefore the lower the portfolio risk and the greater the chance to experience higher portfolio return. When $\rho_{yx} = -1$, an investor may achieve zero portfolio risk ($\sigma^2_P = 0$) and high portfolio return.

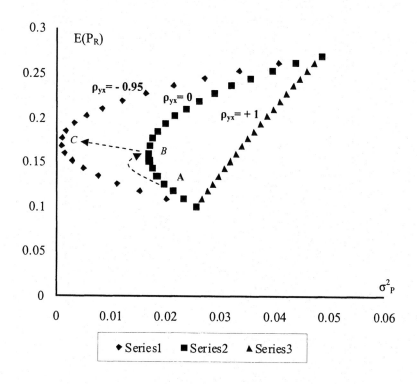

Question 13. Consider a manufacturer of a small commuter plane that can accommodate 30 passengers or less. A market survey of the passengers using this type of plane yield the following (relative frequency) probability (P) distribution of volume of luggage (X) measured in cubic feet:

X	0	1	2	3	4	5	6	7
P	0.01	0.07	0.18	0.34	0.24	0.12	0.03	0.01

Compute the expected value, variance, and standard deviation. Forecast the levels of reasonable space needed for luggage for a full load of 30 passengers.

Answers for #13:

$$E(X) = 0(0.01) + 1(0.07) + 2(0.18) + 3(0.34) + 4(0.24) + 5(0.12) + 6(0.03) + 7(0.01) = 3.26$$

$$\sigma_X^2 = VAR(X) = (0.01)(0-3.26)^2 + (0.07)(1-3.26)^2 + (0.18)(2-3.26)^2 +$$
$$(0.34)(3-3.26)^2 + (0.24)(4-3.26)^2 + (0.12)(5-3.26)^2 + (0.03)(6-3.26)^2 +$$
$$(0.01)(7-3.26)^2 = 1.632$$

$$\sigma_X = \sqrt{1.632} = 1.278.$$

In forecasting the levels of reasonable space needed for luggage for a full load of 30 passengers, one must consider the sum of i future events over time (T). The sum of $i=1$ to n future independent events has:

$$T = X_1 + X_2 + \ldots + X_n$$

and

$$E(T) = \mu_T = nE(X)$$

$$\sigma_T = \sqrt{n\sigma_X^2}.$$

Thus,

$$\mu_T = 30(3.26) = 97.8 \text{ cubic feet}$$

$$\sigma_T = \sqrt{30(1.632)} = 6.99 = \pm 7 \text{ cubic feet.}$$

Hence, we can define an interval for the total volume of 30 passengers by combining both the mean (μ_T) and standard deviation (σ_T) as $\mu_T \pm \sigma_T$:

$$97.8 \pm 7 = 90.8 \leq T \leq 104.8 \text{ cubic feet or,}$$

(90.8, 104.8) cubic feet.

Note: In the next Chapter, we will elaborate upon, and formalize, the concept of interval estimation in the context of confidence intervals.

Chapter 3

SAMPLING DISTRIBUTIONS AND INTERVAL ESTIMATION

"The process of inferring the values of unknown population parameters from known sample statistics is called estimation"

3.1 Sampling Distributions

3.1.1 Sampling Distribution of the Sample Mean

When a sample statistic is used to estimate a population parameter, it will usually be in error by some amount. After all, a statistic is a random variable that fluctuates from sample to sample. Consequently, in estimating a parameter, one should give a bound on the error of the estimate. To do this, one must know the probability distribution of the sample statistic. This is called the *sampling distribution of the statistic*. Often the sampling distribution of interest pertains to the sample mean. The basic concept of a sampling distribution of the sample mean is introduced in the following illustration.

Let a population X consist of a die's possibilities 1, 2, 3, 4, 5, and 6. Each outcome has a probability of occurrence equal to 1/6, and therefore the population's probability distribution is uniform as it may be seen in Figure 3.1.

Figure 3.1

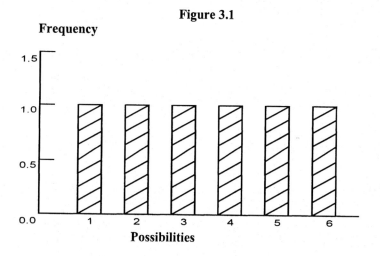

The mean, variance, and standard deviation of the die population are $\mu_x = 3.5$, $\sigma_x^2 = 2.92$, and $\sigma_x = 1.71$.

To illustrate, consider all possible samples of n = 2 that may be derived with replacement from the die population (Table 3.1), their probability distribution (Table 3.2), and their probability histogram (Figure 3.2):

Table 3.1
All Samples of n = 2 from Population X

SAMPLES/SAMPLE SPACE	SAMPLE MEANS
1,1 1,2 1,3 1,4 1,5 1,6	1.0 1.5 2.0 2.5 3.0 3.5
2,1 2,2 2,3 2,4 2,5 2,6	1.5 2.0 2.5 3.0 3.5 4.0
3,1 3,2 3,3 3,4 3,5 3,6	2.0 2.5 3.0 3.5 4.0 4.5
4,1 4,2 4,3 4,4 4,5 4,6	2.5 3.0 3.5 4.0 4.5 5.0
5,1 5,2 5,3 5,4 5,5 5,6	3.0 3.5 4.0 4.5 5.0 5.5
6,1 6,2 6,3 6,4 6,5 6,6	3.5 4.0 4.5 5.0 5.5 6.0

Table 3.2
Probability Distribution for Sample Means
(*Sampling Distribution of the Mean*)

SAMPLE MEANS	FREQUENCY (f)	PROBABILITY (P)
1.0	1	1/36
1.5	2	2/36
2.0	3	3/36
2.5	4	4/36
3.0	5	5/36
3.5	6	6/36
4.0	5	5/36
4.5	4	4/36
5.0	3	3/36
5.5	2	2/36
6.0	1	1/36
	36	1

Figure 3.2

Frequency

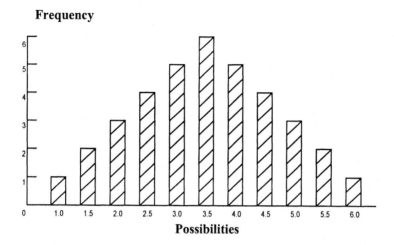

Possibilities

Notice that the sampling distribution of the mean (or the sampling distribution of $\overline{X}$) has a shape that resembles a normal distribution, even though the distribution of the parent population, given in Figure 3.1, is rectangular in shape. Computing the mean ($\mu_{\overline{X}}$), variance ($\sigma_{\overline{X}}^2$), and standard deviation ($\sigma_{\overline{X}}$) of the sampling distribution of the mean we have:

$$\mu_{\overline{X}} = 3.5, \ \sigma_{\overline{X}}^2 = 1.46, \text{ and } \sigma_{\overline{X}} = 1.21.$$

$\sigma_{\overline{X}}$ is often referred to as the *standard error of the mean*.

The mean of the sampling distribution of the mean is equal to the mean of the population. This is always true whenever *all possible samples* from a population are considered. Considering all possible samples from large or infinite populations is impossible. Experience has shown though that, regardless of population size, sampling distributions of the mean generate means which are close to the means of populations.

Summarizing: For random samples of size n selected from a population with mean μ and standard deviation σ, the sampling distribution of the sample mean $\overline{X}$ has a:

1. Mean of $\mu_{\overline{X}} = \mu$, the population mean;

2. Standard Deviation of : $\sigma_{\overline{X}} = \dfrac{\sigma}{\sqrt{n}}$

when n < 0.05N or, N = ∞ or, sampling occurs with replacement or, without replacement and N is large relative to n;

$$\sigma_{\overline{X}} = \left(\dfrac{\sigma}{\sqrt{n}}\right)\left[\dfrac{\sqrt{(N-n)}}{\sqrt{(N-1)}}\right]$$

when n ≥ 0.05N or, sampling occurs without replacement and N is small relative to n.

A major consequence of the above statement is that often a normal probability distribution can be used to find probabilities concerning a sample mean. The statement says that if the shape of the parent population is normal, then the sampling distribution of the sample mean will also be normal. But more remarkably it says that even if the population is not normal, the sampling distribution of the sample mean will still be approximately normal, provided that the sample size n is large. This last result is one of the most important and frequently cited theorems in statistics. It was introduced by Pierre Simon Laplace in 1810 and it is called the *Central Limit Theorem*. This theorem is the basis for inferential statistics. It may be stated as follows:

1.   For random samples selected from a population of *unknown shape*, the sampling distribution of the sample mean $\overline{X}$ is approximately normal when the sample size n is large, or when   30 ≤ n < 0.05N.

2.   If the parent population is normal, the sampling distribution of the sample mean is normal, regardless of sample size n.

For many populations the Central Limit Theorem is still applicable for much smaller than 30 sample sizes. This point has already been illustrated with the die example, where the sample size n was 2. Notice also that $\sigma_{\overline{X}}$ is always less than σ, which implies that the distribution of sample means will always vary less than the population from which the samples are selected.

3.1.2   Sampling Distribution of the Sample Proportion

Researchers are frequently presented with estimates of population proportions or, equivalently, percentages. To estimate a population proportion, we need to consider the problem of estimating the probability of a success or failure in a binomial experiment or, proportion parameter ϕ (read as "phi"). In a binomial experiment outcomes are classified under "success" or "failure" where

ϕ = proportion of successes and,

q = 1-ϕ = proportion of failures.

A point estimate of the population proportion ϕ is obtained by computing the corresponding sample proportion. The sample proportion is denoted by p and equals x/n where x is the number of successes or failures.

The Sampling Distribution of the Sample Proportion (p) may be summarized as follows:

1. The mean of p is $\mu_p = \phi$, the population proportion;

2. The standard deviation of p is: $\sigma_p = \sqrt{\left(\dfrac{\phi q}{n}\right)}$, where q = 1-$\phi$,

when n < 0.05N or, N = ∞ or, sampling occurs with replacement or, without replacement and N is large relative to n.

$$\sigma_p = \left[\sqrt{\left(\frac{\phi q}{n}\right)}\right]\left[\frac{\sqrt{(N-n)}}{\sqrt{(N-1)}}\right]$$

when n $\geq$ 0.05N or, sampling occurs without replacement and N is small relative to n.

Similarly, as in the case of the Sampling Distribution of the Mean, the Central Limit Theorem associated with the sample proportion states that:

The shape of the sampling distribution of the sample proportion is approximately normal for large samples or, provided that np $\geq$ 5 and n(1-p) $\geq$ 5.

Important Note: Notice the differences in notation!
SDM = Sampling Distribution of the Mean and,
SDP = Sampling Distribution of the Sample Proportion

Table 3.3

	Mean	Variance	Standard Deviation
Population	μ_X	σ_X^2	σ_X
SDM	$\mu_{\overline{X}}$	$\sigma_{\overline{X}}^2$	$\sigma_{\overline{X}}$
Sample	$\overline{X}$	s^2	s
SDP	μ_p	σ_p^2	σ_p

Before we continue consider an illustration of both the sampling distribution of the mean and the sampling distribution of the sample proportion.

Suppose we are concerned with the life expectancy differential (female life expectancy minus male life expectancy) of the different countries of the world. For simplicity, let the population consist of N = 5 countries. Life expectancy differential (LED) and state of development (SOD), where D = developed and U = developing, are reported in Table 3.4.

Table 3.4

COUNTRY	LED	SOD
P	2	D
Q	5	U
R	4	U
S	3	D
T	6	U

The population summary measures for LED and SOD are:

LED: $\mu = 4$, and $\sigma = 1.414$.

SOD: $\phi_U = 3/5 = 0.6$.

Consider now all possible samples of n = 3 taken without replacement from this population along with the summary measures of $\overline{X}$, and p_U. [To determine the number of samples, in this case, compute $C_{N,n} = N!/n!(N-n)!$. Thus, $C_{5,3} = 5!/3!(5-3)! = 10$.] The results are summarized in Table 3.5.

Table 3.5

SAMPLE	$\overline{X}$	P_U
PQR	3.67	0.67
PQS	3.33	0.33
PQT	4.33	0.67
PRS	3.00	0.33
PRT	4.00	0.67
PST	3.67	0.33
QRS	4.00	0.67
QRT	5.00	1.00
QST	4.67	0.67
RST	4.33	0.67

Finally, Table 3.6 summarizes the sampling distributions of the mean and of the sample proportion as well as their summary measures and histograms. (P = probability).

Table 3.6

Sampling Distribution of Mean		Sampling Distribution of Sample Proportion	
$\overline{X}$	$P_{\overline{X}}$	P_U	P_{P_U}
3.00	1/10		
3.33	1/10		
3.67	2/10	0.33	3/10
4.00	2/10	0.67	6/10
4.33	2/10	1.00	1/10
4.67	1/10		
5.00	1/10		
$\mu_{\overline{X}} = 4$ $\sigma_{\overline{X}} = 0.557$	10/10 = 1	$\mu_{P_U} = 0.6$ $\sigma_{P_U} = 0.2$	10/10 = 1

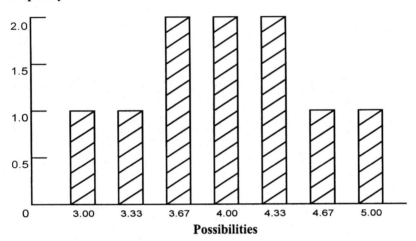

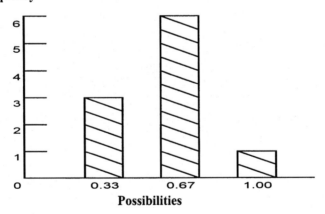

Thus:

$$\mu_{\overline{X}} = \mu = 4,$$

$$\sigma_{\overline{X}} = \left(\frac{\sigma}{\sqrt{n}}\right) \left[\frac{\sqrt{(N-n)}}{\sqrt{(N-1)}}\right] = 0.577,$$

$$\mu_{P_U} = \phi_U = 0.6, \text{ and}$$

$$\sigma_{P_U} = \left[\sqrt{\left(\frac{\phi_U q}{n}\right)}\right] \left[\frac{\sqrt{(N-n)}}{\sqrt{(N-1)}}\right] = 0.2.$$

Additionally, as the central limit theorems predict, both sampling distributions resemble normal distributions.

Important Note: For every population parameter there is one best estimator or statistic. Each statistic has its own sampling distribution. The form of these sampling distributions varies, but in concept they are like the sampling distributions described above.

(Descriptions of additional sampling distributions may be found in: *Statistics for Business and Economics*, by H. Kohler, Harper Collins, 3rd edition, 1994.)

3.2 Continuous Probability Distributions

Before we continue with applications we would like to briefly describe *Continuous Probability Distributions*. As with discrete variables, probability distributions are used in calculating probabilities of continuous variables. However for a continuous random variable we only calculate the probability that the variable will assume a value within a specified interval.

Moreover, this probability is obtained by determining the area under the graph of the random variable's probability distribution. For instance, if X is a continuous random variable with probability distribution f(X), then $P(a \leq X \leq b)$, the probability that x will assume a value within the interval from x = a to x = b, equals the area under the graph of f(X) from x = a to b.

3.2.1   The Normal Distribution

Normal probability distributions are the most important of the continuous distributions. In addition to their vital role in the theory of statistics, normal distributions serve as mathematical models for many applications. There are an infinite number of different normal distributions (the sampling distribution of the mean being one), and a particular one is determined by two parameters: its mean μ, and its standard deviation σ.

The normal probability distribution is usually denoted by X, it is given by the formula,

$$f(X) = \left[\frac{1}{\sigma\sqrt{2\pi}}\right] e^{\frac{-(x-\mu)^2}{2\sigma^2}} \quad \text{for} \quad -\infty < X < +\infty,$$

The Normal Distribution is "Bell-Shaped" in Figure 3.3.

Figure 3.3

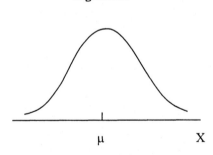

3.2.2   The Standard Normal or Z-Distribution

The normal probability distribution with $\mu = 0$ and $\sigma = 1$ is called *Standard Normal Distribution* and it is denoted by Z. Figure 3.4 shows the correspondence between X (normal distribution) and Z (standard normal).

Figure 3.4

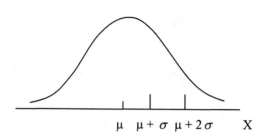

$$\mu \quad \mu + \sigma \quad \mu + 2\sigma \qquad X$$

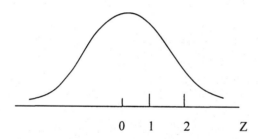

$$0 \quad 1 \quad 2 \qquad Z$$

With only a table of areas for the Standard Normal or Z-distribution, we can determine probabilities for every normal random variable by using the following transformation:

$$Z = \frac{(X - \mu)}{\sigma} \qquad \text{or,}$$

$$Z = \frac{\left(X - \mu_{\overline{X}}\right)}{\sigma_{\overline{X}}} \qquad \text{or,}$$

$$Z = \frac{\left(p - \mu_p\right)}{\sigma_p}$$

Example #1:
Suppose the length of time, X, students need for a test is normally distributed
with a mean of 30 minutes and a standard deviation of 10 minutes. What is the
probability that a student's time will be greater than 55 minutes?

We are looking for the probability that X is greater than 55, or P(X ≥ 55).
Substituting X = 55, μ = 30, and σ = 10 into Z we have:

$$Z = \frac{(55 - 30)}{10} = 2.5$$

From the Z-table (see Appendix III – Standardized Normal Distribution), which
shows the Standard Normal Curve Areas to the right of zero, we see that the area
between zero and 2.5 is 0.4938.
Thus,

P(X ≥ 55) = P(Z ≥ 2.5) = 0.5-0.4938 = 0.0062.

See Figure 3.5.

Figure 3.5

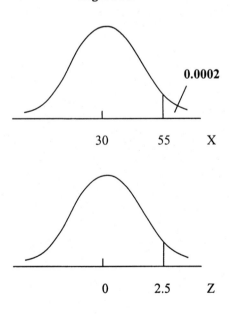

Example #2:

In the previous example, what is the probability that a student's time will be less than 22 minutes? We are looking for the probability that X is less than 22, or $P(X \leq 22)$. Substituting $X = 22$, $\mu = 30$, and $\sigma = 10$ into Z we have:

$$Z = \frac{(22 - 30)}{10} = -0.8$$

Thus,

$$P(X \leq 22) = P(Z \leq -0.8) = 0.5 - 0.2881 = 0.2119.$$

See Figure 3.6.

Figure 3.6

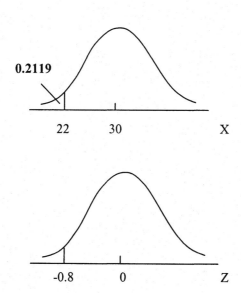

Example #3:

In the above example, what is the probability that a student's time will be between 35 and 50 minutes? We are looking for $P(35 \leq X \leq 50)$. In this case we have two X values: $X_1 = 35$ and $X_2 = 50$. Therefore the corresponding Z values are:

$$Z_1 = \frac{(5-30)}{10} = 0.5 \text{ and } Z_2 = \frac{(50-30)}{10} = 2.$$

Thus,

$$P(35 \leq X \leq 50) = P(0.5 \leq Z \leq 2) = P(Z_2) - P(Z_1) =$$
$$0.4772 - 0.1915 = 0.2857$$

See Figure 3.7.

Figure 3.7

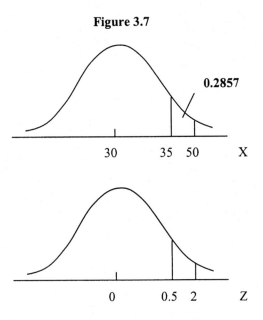

Example #4:
The upper 3% of scores in a Statistics exam will receive As. What is the lowest score to be designated an A? The scores are normally distributed with a mean of 83 and a standard deviation of 5.

See Figure 3.8. Here we know the area under the X curve (0.03), and we do not know the number that corresponds to X_L= lowest score to be designated an A. The Z number that corresponds to X_L is Z_L. The area to the right of Z_L is 0.03 and the area between 0 and Z_L is 0.47. Thus, from the Z table we see that 0.47 corresponds approximately to $Z_L = 1.88$. (See the following graph.)

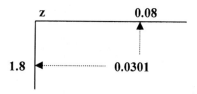

Therefore,

$$Z = \frac{(X - \mu)}{\sigma} \Rightarrow Z_L = \frac{(X_L - \mu)}{\sigma} \Rightarrow 1.88 = \frac{(X_L - 83)}{5} \Rightarrow X_L = 92.4.$$

Figure 3.8

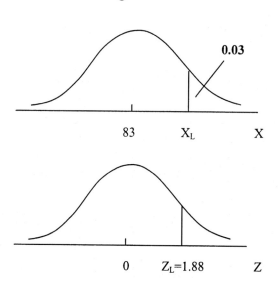

Example #5:
In the previous example, if the upper 20% of scores is to receive As and Bs, what is the lowest score to be designated a B?

$$Z = \frac{(X - \mu)}{\sigma} \Rightarrow Z_L = \frac{(X_L - \mu)}{\sigma} \Rightarrow 0.84 = \frac{(X_L - 83)}{5} \Rightarrow X_L = 87.2$$

Example #6:
A population has a normal distribution with a mean of 98 and a standard deviation of 12. If 9 measurements are randomly selected from this population, what is
the probability that the sample mean will exceed 104?

Here we are concerned with the computing of a probability under the sampling
distribution of the sample mean X. We want to compute the P($\overline{X} \geq 104$). Therefore,

$$Z = \frac{(\overline{X} - \mu_{\overline{x}})}{\sigma_{\overline{x}}} = \frac{(\overline{X} - \mu)}{\left(\dfrac{\sigma}{\sqrt{n}}\right)} = \frac{(104 - 98)}{\left(\dfrac{12}{\sqrt{9}}\right)} = 1.5$$

Thus, P($\overline{X} \geq 104$) = P(Z $\geq$ 1.5) = 0.0668. See Figure 3.9.

Figure 3.9

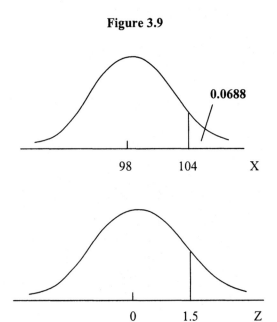

Example #7:

Suppose that 65% of all voters are expected to vote for candidate X; thus, $\phi = 0.65$. If $\mu_p = 0.65$ and $\sigma_p = 0.06$, what is the probability that a sample of n = 200 would yield a percentage greater than 85?

Because $\phi n = 0.65(200) = 130$ and $(1-\phi)n = 70$ the sampling distribution of the sample proportion can be assumed to be normal. Therefore, we can use the Z-distribution.

See Figure 3.10.

Figure 3.10

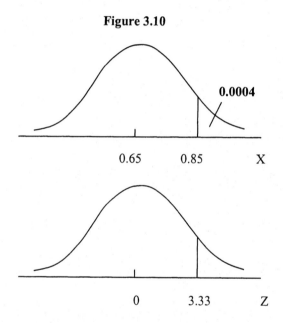

$$Z = \frac{(p - \mu_p)}{\sigma_p} = \frac{(0.85 - 0.65)}{0.06} = 3.33$$

Thus, $P(p \geq 0.85) = P(Z \geq 3.33) = 0.5 - 0.4996 = 0.0004$

3.2.3 The t-Distribution

A probability distribution that resembles the Z-distribution is the Student's *t-distribution* (or Gosset distribution). It has a greater spread than the Z-distribution, and each particular curve is determined by a single parameter, called *degrees of freedom* (df).

To understand the meaning of df consider the following illustration: If the sum of four numbers is 20, various combinations of three numbers can be written down, but the fourth number is restricted. If you select 7, 4, and 1 as three of the numbers, the fourth number must be 8 so that the sum of all is 20. Because of this restriction, it is said that "1 degree of freedom is lost."

df = n-k, where k=number of parameters under investigation.

We use the t-distribution when df < 30 and the parent population is normal. We can determine probabilities for every normal random variable by using a table of areas for the t-distribution (see Appendix III – Percentile of the *t* Distribution) based on the following transformation:

$$t = \frac{(\overline{X} - \mu)}{\left(\dfrac{s}{\sqrt{n}}\right)}, \text{ where s = sample's standard deviation.}$$

When df ≥ 30 the t distribution is approximately equal to the Z distribution. See Figure 3.11.

Figure 3.11

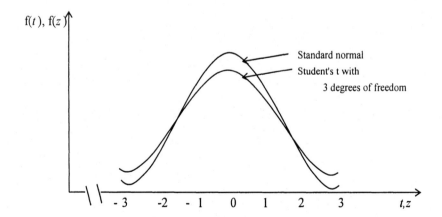

Example #8:

For the t-distribution with df = 10, find the number c such that

$$P(t \geq c) = 0.05.$$

From the t-table we find $P(t \geq 1.812) = 0.05$ and thus $t_{0.05} = 1.812$.

See Figure 3.12.

Figure 3.12

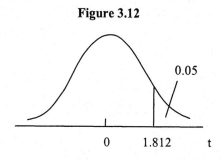

0.05

0   1.812   t

Example #9:

For the t-distribution with df = 29, find the number c such that

$$P(t \leq c) = 0.025.$$

From the t-table we find $P(t \leq -2.045) = 0.025$ and thus $t_{0.025} = -2.045$.

See Figure 3.13.

Figure 3.13

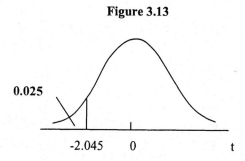

0.025

-2.045   0   t

3.3 Confidence Interval for the Mean of the Population

We learned above that the SDM is normally distributed, if the parent population is normally distributed regardless of sample size, and approximately normally distributed, if $30 \leq n < 0.05N$ regardless of the shape of the parent population.

Correspondingly, $Z = \dfrac{(\overline{X} - \mu_{\overline{x}})}{\sigma_{\overline{x}}}$ is standard normally (or approximately

normally) distributed. The probability that Z falls between any two limits $-Z_{\alpha/2}$ and $Z_{\alpha/2}$, where $0 \leq \alpha \leq 1$, is

$$P[-Z_{\alpha/2} \leq Z \leq Z_{\alpha/2}] = 1 - \alpha.$$

Figure 3.14 shows the correspondence between Z and $\overline{X}$.

Figure 3.14

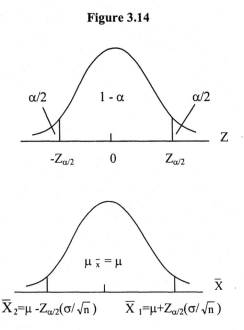

Thus, there is a $(1-\alpha)$ *pre-sampling* probability that $\overline{X}$ will fall in the interval

$$\mu \pm Z_{\alpha/2}\left(\frac{\sigma}{\sqrt{n}}\right) \text{, or}$$

$$P\left[\mu - Z_{\alpha/2}\left(\frac{\sigma}{\sqrt{n}}\right) \leq \overline{X} \leq \mu + Z_{\alpha/2}\left(\frac{\sigma}{\sqrt{n}}\right)\right] = 1-\alpha.$$

For example, at $\alpha = 0.05$, $1-\alpha = 0.95$, $\alpha/2 = 0.025$, $Z_{0.025} = 1.96$, and $\overline{X}_1 = \mu + 1.96\left(\frac{\sigma}{\sqrt{n}}\right)$, $\overline{X}_2 = \mu - 1.96\left(\frac{\sigma}{\sqrt{n}}\right)$. Hence, there is a 0.95 pre-sampling probability that $\overline{X}$ will fall in the interval $\mu \pm 1.96\left(\frac{\sigma}{\sqrt{n}}\right)$.

Consider now the equation $P\ [-Z_{\alpha/2} \leq Z \leq Z_{\alpha/2}] = 1-\alpha$ that we discussed above. Since $Z = \dfrac{\left(\overline{X} - \mu_{\overline{X}}\right)}{\sigma_{\overline{X}}}$, where $\mu_{\overline{X}} = \mu$ and $\sigma_{\overline{X}} = \left(\dfrac{\sigma}{\sqrt{n}}\right)$, we have,

$$P\left[-Z_{\alpha/2} \leq \frac{\left(\overline{X} - \mu\right)}{\left(\dfrac{\sigma}{\sqrt{n}}\right)} \leq +Z_{\alpha/2}\right] = 1-\alpha.$$

or,

$$P\left[-Z_{\alpha/2}\left(\frac{\sigma}{\sqrt{n}}\right) \leq \left(\overline{X} - \mu\right) \leq Z_{\alpha/2}\left(\frac{\sigma}{\sqrt{n}}\right)\right] = 1-\alpha.$$

or,

$$P\left[-Z_{\alpha/2}\left(\frac{\sigma}{\sqrt{n}}\right) - \overline{X} \leq -\mu \leq Z_{\alpha/2}\left(\frac{\sigma}{\sqrt{n}}\right) - \overline{X}\right] = 1-\alpha.$$

or,

$$\mu \geq -Z_{\alpha/2}\left(\frac{\sigma}{\sqrt{n}}\right) + \overline{X}$$

$$P\left[-Z_{\alpha/2}\left(\frac{\sigma}{\sqrt{n}}\right) - \overline{X} \leq -\mu \leq Z_{\alpha/2}\left(\frac{\sigma}{\sqrt{n}}\right) - \overline{X}\right] = 1-\alpha.$$

$$Z_{\alpha/2}\left(\frac{\sigma}{\sqrt{n}}\right) + \overline{X} \geq \mu$$

$$P\left[\overline{X} - Z_{\alpha/2}\left(\frac{\sigma}{\sqrt{n}}\right) \leq \mu \leq \overline{X} + Z_{\alpha/2}\left(\frac{\sigma}{\sqrt{n}}\right)\right] = 1-\alpha.$$

This last equation implies that there is a $(1-\alpha)$ *post-sampling* probability that μ will fall in the interval $\overline{X} \pm Z_{\alpha/2}\left(\frac{\sigma}{\sqrt{n}}\right)$.

Because the $\overline{X}$ numbers are means of samples, then, given $1-\alpha$, one may construct as many $\overline{X} \pm Z_{\alpha/2}\left(\frac{\sigma}{\sqrt{n}}\right)$ intervals as there are samples in the SDM. It may be shown that $[(1-\alpha)100]\%$ of these intervals will contain μ, the unknown population mean.

The interval $\overline{X} \pm Z_{\alpha/2}\left(\frac{\sigma}{\sqrt{n}}\right)$ is a $[(1-\alpha)100]\%$ confidence interval for the unknown population mean μ, where $(1-\alpha)$ is the confidence level. For instance, if

$1-\alpha = 0.95$ then we may claim that we are 95% confident that μ lies in the interval:

$$\overline{X} \pm 1.96\left(\frac{\sigma}{\sqrt{n}}\right).$$

Example #1:
Consider the SDM reported in Table 3.2. The mean of the parent population ($\mu = 3.5$) is of course known but, let us pretend that it is unknown and that we desire to estimate its value with 95% confidence. Thus, from

$$P\left[\overline{X} - Z_{\alpha/2}\left(\frac{\sigma}{\sqrt{n}}\right) \leq \mu \leq \overline{X} + Z_{\alpha/2}\left(\frac{\sigma}{\sqrt{n}}\right)\right] = 1-\alpha.$$

where $1-\alpha = 0.95$, $Z_{\alpha/2} = 1.96$, $\sigma = 1.71$, and $n = 2$, we have the 95% confidence interval estimator for μ, or

$$P[\overline{X} - 2.36994 \leq \mu \leq \overline{X} + 2.36994] = 0.95$$

It means that, if we draw all possible samples of size 2 from the parent population, approximately 95% of the values of $\overline{X}$ will be such that μ would lie somewhere between $\overline{X}$ - 2.36994 and $\overline{X}$ + 2.36994, and approximately 5% of the values of $\overline{X}$ will produce intervals that would not include μ.

The table on the following page reports the results. Notice that only the intervals marked by an asterisk (*) do not include the value of the true population mean. Thus, 34 out of 36, or 94.44%, of the intervals include the mean and only 2 out of 36, or 5.55%, intervals do not include it.

SAMPLE	$\overline{X}$	$\overline{X} - 2.36994$	$\overline{X} + 2.36994$
1	1.000	-1.370	3.370*
2	1.500	-0.870	3.870
3	1.500	-0.870	3.870
4	2.000	-0.370	4.370
5	2.000	-0.370	4.370
6	2.000	-0.370	4.370
7	2.500	0.130	4.870
8	2.500	0.130	4.870
9	2.500	0.130	4.870
10	2.500	0.130	4.870
11	3.000	0.630	5.370
12	3.000	0.630	5.370
13	3.000	0.630	5.370
14	3.000	0.630	5.370
15	3.000	0.630	5.370
16	3.500	1.130	5.870
17	3.500	1.130	5.870
18	3.500	1.130	5.870
19	3.500	1.130	5.870
20	3.500	1.130	5.870
21	3.500	1.130	5.870
22	4.000	1.630	6.370
23	4.000	1.630	6.370
24	4.000	1.630	6.370
25	4.000	1.630	6.370
26	4.000	1.630	6.370
27	4.500	2.130	6.870
28	4.500	2.130	6.870
29	4.500	2.130	6.870
30	4.500	2.130	6.870
31	5.000	2.630	7.370
32	5.000	2.630	7.370
33	5.000	2.630	7.370
34	5.500	3.130	7.870
35	5.500	3.130	7.870
36	6.000	3.630	8.370*

Example #2:
Consider the previous example: What are the 99% and 90% confidence intervals for the mean of the population?

With $\sigma = 1.71$, and n = 2, we are looking for:

$$P\left[\overline{X} - Z_{\alpha/2}\left(\frac{\sigma}{\sqrt{n}}\right) \leq \mu \leq \overline{X} + Z_{\alpha/2}\left(\frac{\sigma}{\sqrt{n}}\right)\right] = 1-\alpha.$$

With 99% confidence:
$1-\alpha = 0.99$, $Z_{\alpha/2} = 2.575$,
and $P[\overline{X} - 3.1136 \leq \mu \leq \overline{X} + 3.1136] = 0.99$;

With 90% confidence:
$1-\alpha = 0.90$, $Z_{\alpha/2} = 1.645$,
and $P[\overline{X} - 1.9891 \leq \mu \leq \overline{X} + 1.9891] = 0.90$;

Important Note: Notice that the confidence level and the size of the interval move in the same direction: *The higher the confidence level the wider the confidence interval.* Notice also that given a certain confidence level, *the larger the sample size, n, the narrower the confidence interval.*

Example #3:
Previous studies have established that the standard deviation of population X is 23. A random sample (derived without replacement) of size 144 from this population has a mean of 45. What are the 99% and 95% confidence intervals for the mean of the population?

With $\sigma = 23$, n = 144, and $\overline{X} = 45$, we are looking for

$$P\left[\overline{X} - Z_{\alpha/2}\left(\frac{\sigma}{\sqrt{n}}\right) \leq \mu \leq \overline{X} + Z_{\alpha/2}\left(\frac{\sigma}{\sqrt{n}}\right)\right] = 1-\alpha.$$

With 99% confidence:
$1-\alpha = 0.99$, $Z_{\alpha/2} = 2.575$, and $P(40.0648 \leq \mu \leq 49.9352) = 0.99$,
or $\mu = 45 \pm 4.9352$ with 99% confidence;

With 95% confidence:
$1-\alpha = 0.95$, $Z_{\alpha/2} = 1.96$, and $P(41.2435 \leq \mu \leq 48.7565) = 0.95$,

or $\mu = 45 \pm 3.7565$ with 95% confidence.

In real world applications not only the population mean is unknown but also the population variance. If the size of the sample falls in $30 \le n < 0.05N$, σ may be approximated by the standard deviation, s, of the sample, or $\sigma_x \approx \left(\dfrac{s}{\sqrt{n}} \right)$.

Example #4:
A random sample of 49 observations (derived with replacement) has a mean of 25 and a standard deviation of 4. Compute the 99% confidence interval about the mean of the population.

With $s = 4$, $n = 49$, and $\overline{X} = 25$, we are looking for:

$$P\left[\overline{X} - Z_{\alpha/2}\left(\frac{s}{\sqrt{n}} \right) \le \mu \le \overline{X} + Z_{\alpha/2}\left(\frac{s}{\sqrt{n}} \right) \right] = 1\text{-}\alpha.$$

Since, $1 - \alpha = 0.99$, and $Z_{\alpha/2} = 2.575$,
we have $P(23.5286 \le \mu \le 26.4714) = 0.99$,
or $\mu = 25 \pm 1.4714$ with 99% confidence.

When $n < 30$, and the parent population is normal, we use the t-distribution to compute confidence intervals as follows:

$$P\left[\overline{X} - t_{\alpha/2}\left(\frac{s}{\sqrt{n}} \right) \le \mu \le \overline{X} + t_{\alpha/2}\left(\frac{s}{\sqrt{n}} \right) \right] = 1\text{-}\alpha.$$

Important Note: In practice we use the t-distribution to make inferences based on small samples about all types of populations, except those which are highly skewed.

Example #5:
A random sample of 21 observations (derived from a normal population) has a mean of 25 and a standard deviation of 4. Compute the 99% confidence interval about the mean of the population.

$$df = n\text{-}1 = 21\text{-}1 = 20, \overline{X} = 25, \text{ and } s = 4.$$

From the t table $t_{\alpha/2} = 2.845$. Therefore,

$$P\left[\overline{X} - t_{\alpha/2}\left(\frac{s}{\sqrt{n}}\right) \leq \mu \leq \overline{X} + t_{\alpha/2}\left(\frac{s}{\sqrt{n}}\right)\right] = 1\text{-}\alpha.$$

$$P(22.7934 \leq \mu \leq 27.2066) = 0.99$$

Example #6:
A random sample has been derived from a normal population. The observations are:

$$12, 15, 21, 22, 14, 15, 17, 19, 17, 10, 9$$

Compute the 95% confidence interval about the mean of the population. From the sample data we can compute the sample mean and sample standard deviation:

$$\overline{X} = \frac{\Sigma X}{n} = 15.545$$

$$s = \sqrt{\frac{\Sigma (X - \overline{X})^2}{n-1}} = 4.204$$

$$df = n\text{-}1 = 11\text{-}1 = 10.$$

From the t table $t_{\alpha/2} = 2.228$. Therefore,

$$P\left[\overline{X} - t_{\alpha/2}\left(\frac{s}{\sqrt{n}}\right) \leq \mu \leq \overline{X} + t_{\alpha/2}\left(\frac{s}{\sqrt{n}}\right)\right] = 1\text{-}\alpha.$$

$$P(12.7209 \leq \mu \leq 17.7730) = 0.95,$$
or $\mu = 15.545 \pm 2.8241$ with 95% confidence.

3.4 Confidence Interval for the Population Proportion

With a large sample, or $np \geq 5$, and $n(1-p) \geq 5$ (where p = sample propor-

tion), $\sigma_p = \sqrt{\dfrac{\phi q}{n}} \approx \sqrt{\dfrac{p(1-p)}{n}}$. Thus, the confidence interval for, ϕ, the popula-

tion proportion is:

$$P\left[(p - (Z_{\alpha/2})\sigma_p) \leq \phi \leq (p + (Z_{\alpha/2})\sigma_p)\right] = 1 - \alpha$$

Example#1:
Consider a sample of size $n = 120$. Let $x = 75$, where x = "tested positive."
Compute the 95% confidence interval for the proportion of the population.

$p = 75/120 = 0.625$, and $Z_{\alpha/2} = 1.96$.

Therefore,

$P(0.538 \leq \phi \leq 0.712) = 0.95$,
or $\phi = 0.625 \pm 0.087$, with 95% confidence.

Example#2:
237 out of 500 individuals sampled said that they would vote for candidate Y.
Compute the 99% confidence interval for the proportion of the population.

$p = 237/500 = 0.474$, and $Z_{\alpha/2} = 2.575$.

Therefore,

$P(0.4165 \leq \phi \leq 0.5315) = 0.99$, or
$\phi = 0.474 \pm 0.0575$, with 99% confidence.

3.5 Confidence Interval for the Variance of the Population

In some situations, a researcher may be concerned with estimating the variance, σ^2, of a population. It is important because it measures the variability in a
population, which in turn may affect the precision of some process.

The Confidence Interval for σ^2 is based on the chi-square (χ^2) distribution
and it is computed as follows:

$$P\left[\frac{(n-1)s^2}{\chi_L^2} \leq \sigma^2 \leq \frac{(n-1)s^2}{\chi_U^2}\right] = 1 - \alpha$$

where n is the sample size, and s^2 the sample variance. $U = \alpha/2$ and $L = 1 - (\alpha/2)$ are chi-square values whose *right-hand areas* are based on the chi-square (χ^2) distribution with df = n-1 (see Appendix III - Percentile of the χ^2 Distribution Table). The χ^2 distribution ranges in values from 0 to $+ \infty$. It is skewed to the right, and its skewness decreases with increasing numbers of degrees of freedom. As df approaches $+ \infty$, the shape of the χ^2 distribution approaches the shape of a normal distribution (See Figure 3.15).

Figure 3.15

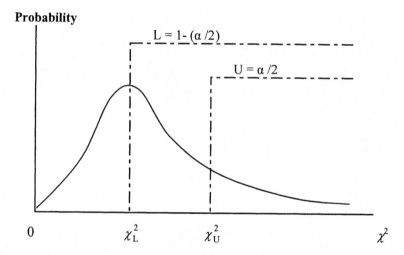

Example #1:
A sample may be summarized by the following information:

$$n = 10, s^2 = 26.7.$$

Compute the 95% confidence interval for the population variance and standard deviation.

$$1 - \alpha = 0.95, df = 9, U = 0.025, L = 0.975,$$

from the χ^2 table, $\chi_L^2 = 19.02$, and $\chi_U^2 = 2.7$.

Therefore,

$$P(12.63 \le \sigma^2 \le 89) = 0.95, \text{ and } P(3.6 \le \sigma^2 \le 9.4) = 0.95.$$

Thus, with 95% confidence the σ^2 varies between 12.63 and 89, and σ between 3.6 and 9.4.

3.6 Confidence Intervals for *Two* Population Parameters

The methodology of confidence intervals for one population parameter may be extended to two population parameters.

a.   The Confidence Interval for $(\mu_1 - \mu_2)$ when $30 \le n_1 < 0.05N_1$, and $30 \le n_2 < 0.05N_2$ is:

$$(\overline{X}_1 - \overline{X}_2) \pm Z_{\alpha/2} \sqrt{\frac{\sigma_1^2}{n_1} + \frac{\sigma_2^2}{n_2}}$$

Example #1:
Some of the statistics generated from two different procedures are reported as follows:

Procedure 1	Procedure 2
$n_1 = 100$	$n_2 = 120$
$\overline{X}_1 = 17$	$\overline{X}_2 = 15$
$s_1 = 2$	$s_2 = 3$

Use a 95%, and then a 99% confidence interval to estimate the difference in the means of the populations from which the samples were drawn.

Thus:

With 95% confidence $(17 - 15) \pm 1.96 \sqrt{\dfrac{2^2}{100} + \dfrac{3^2}{120}}$

With 99% confidence $(17 - 15) \pm 2.575\sqrt{\dfrac{2^2}{100} + \dfrac{3^2}{120}}$

Or, the mean of population 1 exceeds the mean of population 2 by 2 ± 0.6646 with 95% confidence, and by 2 ± 0.873 with 99% confidence.

b. The Confidence Interval for $(\mu_1 - \mu_2)$ when $n_1 < 30$ or $n_2 < 30$ are drawn from normal populations, with equal or approximately equal sigmas, and with df $= n_1+n_2-2$ is:

$$(\overline{X}_1 - \overline{X}_2) \pm t_{\alpha/2}\sqrt{s_p^2\left(\frac{1}{n_1} + \frac{1}{n_2}\right)}$$

where

$$s_p^2 = \frac{(n_1 - 1)s_1^2 + (n_2 - 1)s_2^2}{n_1 + n_2 - 2}$$

Example #2:
Data for samples 1 and 2, drawn from normal populations, are reported as follows:

Sample 1	10	12	15	16	17	9	8	3	5	11
Sample 2	8	16.5	2.5	4.5	8	6	12	15	13	10

Use a 90%, and then a 95% confidence interval to estimate the difference in the means of the populations from which the samples were drawn. From the data, we have:

Procedure 1	Procedure 2
$n_1 = 10$	$n_2 = 10$
$\overline{X}_1 = 10.6$	$\overline{X}_2 = 9.55$
$s_1 = 4.6$	$s_2 = 4.579$

df $= n_1+n_2-2 = 18$, thus:

With 90% confidence $(10.6 - 9.55) \pm 1.734\sqrt{21.064(0.2)}$

With 95% confidence $(10.6 - 9.55) \pm 2.101\sqrt{21.064(0.2)}$

Or, the mean of population 1 exceeds the mean of population 2 by 1.05 ± 3.56 with 90% confidence, and by 1.05 ± 4.31 with 95% confidence.

c.   The Confidence Interval for $(\mu_1 - \mu_2)$ when $n_1 < 30$ or $n_2 < 30$ are drawn from normal populations, with:

$$df = \frac{\left(\dfrac{s_1^2}{n_1} + \dfrac{s_2^2}{n_2}\right)}{\dfrac{\left(\dfrac{s_1^2}{n_1}\right)^2}{n_1 - 1} + \dfrac{\left(\dfrac{s_2^2}{n_1}\right)^2}{n_2 - 1}}$$

is $\left(\overline{X}_1 - \overline{X}_2\right) \pm t_{\alpha/2} \sqrt{\dfrac{s_1^2}{n_1} + \dfrac{s_2^2}{n_2}}$

Example #3:
Data for samples 1 and 2, drawn from normal populations, are reported below:

Sample 1	22	23	18	12	14	9	15	16	14	4	2	16
Sample 2	12	23	24	17	16	50	9	10				

Use a 95%, then a 99% confidence interval to estimate the difference in the means of the populations from which the samples were drawn. From the data we have:

Procedure 1	Procedure 2
$n_1 = 12$	$n_2 = 8$
$\overline{X}_1 = 13.75$	$\overline{X}_2 = 20.125$
$s_1 = 6.341$	$s_2 = 13.282$
$n_1 - 1 = 11$	$n_2 - 1 = 7$

$df = 9.15 \approx 9$, thus:

With 95% confidence $(13.75 - 20.125) \pm 2.365\sqrt{25.401}$

With 99% confidence $(13.75 - 20.125) \pm 3.499\sqrt{25.401}$

Or, the mean of population 1 exceeds the mean of population 2 by 6.375 ± 11.92 with 95% confidence, and by 6.375 ± 17.63 with 99% confidence.

d. The Confidence Interval for two population proportions (ϕ_1 - ϕ_2) when $30 \leq n_1 < 0.05N_1$, and $30 \leq n_2 < 0.05N_2$ is:

$$(p_1 - p_2) \pm Z_{\alpha/2} \sqrt{\frac{p_1(1-p_1)}{n_1} + \frac{p_2(1-p_2)}{n_2}}$$

Example #4:
52 out of 130 women sampled indicated that they would vote for candidate A. 70 out of 200 men sampled indicated that they would vote for the same candidate. Use a 99% confidence interval to estimate the difference in the population proportions from which the samples were drawn. Thus with 99% confidence:

$$(0.4 - 0.35) \pm 2.575 \sqrt{\left[\left(\frac{0.24}{130}\right) + \left(\frac{0.2275}{200}\right)\right]}$$

3.7 Confidence Intervals for Health and Medical Applications

The general concept of sensitivity was discussed in section 1.2.7. However, we can distinguish between sample sensitivity and population sensitivity just as we distinguish such things as the sample mean and the population mean. The sample standard deviation for a given sample sensitivity is:

$$s = \sqrt{\frac{(Sensitivity_{Sample})(1 - Sensitivity_{Sample})}{n}}$$

As discussed in section 3.4, with a large sample, or $np \geq 5$, and $n(1-p) \geq 5$ (where $p = Sensitivity_{Sample}$), the confidence interval for the sensitivity of a population is:

$$P\left[p - (Z_{\alpha/2})(s) \leq Sensitivity_{Population} \leq p + (Z_{\alpha/2})(s)\right] = 1 - \alpha$$

Example #1:
Let us assume that a sample of 79 patients used for a particular study yields a sample sensitivity of 0.557. The sample standard deviation is

$$s = \sqrt{\frac{(0.557)(0.443)}{79}} = 0.056$$

The 95% confidence interval for the sensitivity of the population is:

$$P[\ 0.557 - 1.96(0.056) \leq \text{Sensitivity}_{\text{Population}} \leq 0.557 + 1.96(0.056)\ ] = 0.95$$

$$P[\ 0.44724 \leq \text{Sensitivity}_{\text{Population}} \leq 0.667\ ] = 0.95$$

Thus, in every 95 out of 100 trials, the sensitivity of the population is estimated to be between 0.44724 and 0.667.

3.8 Sample Size

a. Sample size for μ_1

In designing an experiment to estimate a parameter, one of the first things that must be decided is how large the sample size (n) should be.

When the sample mean, with $n \leq 30$, is used to estimate the population mean, the probability is $(1-\alpha)$ that $\overline{X}$ will be different from μ by at most $E = Z_{\alpha/2}\left(\dfrac{\sigma}{\sqrt{n}}\right)$. E is called the *maximum error of the estimate* and equals one-half the width of the confidence interval. E is always fixed by the researcher because it corresponds to how much error the researcher can tolerate. Hence, another name for E is the *Tolerable Error*. Solving for n from equation $E = Z_{\alpha/2}\left(\dfrac{\sigma}{\sqrt{n}}\right)$, we find the *Sample Size Formula When $\overline{X}$ is Used to Estimate μ:*

$$n = \left[\frac{(\sigma)(Z_{\alpha/2})}{E}\right]^2$$

To use this formula, we must have a value for σ. In practice, σ is seldom known and it must be approximated by deriving a *pilot sample* of size n, where

$30 \leq n < 0.05N$, from the population under study. The s of such a pilot sample is approximately equal to the σ of the population. In other words, $\sigma \approx s$.

Example #1:
A pilot sample from a certain population has generated s = 12. If we wish to construct a 99% confidence interval about the mean of the population, what should the sample size be with E = 1.2? With E = 2? Interpret these results.

$$\text{With E} = 1.2, n = \left[\frac{(12)(2.575)}{1.2} \right]^2 \approx 663$$

$$\text{With E} = 2, n = \left[\frac{(12)(2.575)}{2} \right]^2 \approx 239$$

Thus, if we select a sample of 663 (239) observations, the mean of the population would be equal to the mean of the sample plus or minus 1.2 (2) with 99% confidence.

b. Sample size for ϕ

Similarly for ϕ, the population's proportion, $E = Z_{\alpha/2}\left(\sqrt{\frac{\phi q}{n}} \right)$, from which we can compute the *Sample Size Formula when p is used to Estimate ϕ*:

$$n = \phi q \left(\frac{Z_{\alpha/2}}{E} \right)^2$$

where $\phi = 0.5$ with no prior knowledge about ϕ. If prior knowledge about ϕ indicates that it will fall within some range, use the value nearest to 0.5.

Example #2:
Mrs. Y is running for political office. She would like to estimate within 3 percentage points, and with 99% confidence, the proportion of people that would vote for her. How large should the sample size be? What if she wants to be 95% confident? Interpret the results.

Thus, with e = 0.03 and 99% confidence:

$$n = (0.5)(0.5)\left(\frac{2.575}{0.03}\right)^2 \approx 1,842$$

This means that if she selects a sample of 1,842 voters, the proportion of the population that would vote for her, would equal the proportion of the sample that would vote for her plus or minus 0.03.

With the same E but 95% confidence, she needs a smaller sample:

$$n = (0.5)(0.5)\left(\frac{1.96}{0.03}\right)^2 \approx 1,067$$

This means that if she selects a sample of 1,067 voters, the proportion of the population that would vote for her, would equal the proportion of the sample that would vote for her plus or minus 0.03.

Important Note: Note that, given a certain confidence level, n and E are inversely related: a lower E requires a higher n. Whether or not we select a higher or a lower E depends on factors such as: 1) the cost of sampling; 2) the importance of the project under consideration; 3) availability of information; etc. Notice also that, given a certain E, n and the confidence level are directly related: a higher confidence level requires a higher n.

c.   Sample size for $(\mu_1 - \mu_2)$

Similarly, the proper sample size for the difference in population means is found by the following formula:

$$n = \frac{\left(Z_{\alpha/2}\right)^2 \left(\sigma_1^2 + \sigma_2^2\right)}{E^2}$$

Example #3:
Pilot sample 1, from population 1, has generated $s_1 = 12$. Pilot sample 2, from population 2, has generated $s_2 = 18$. If we desire to construct a 99% confidence interval about $(\mu_1 - \mu_2)$, what should the sample sizes be with E=1.5? Interpret the result.

$$n = \frac{(2.575)^2(12^2 + 18^2)}{1.5^2} \approx 1,379$$

1,379 observations should be selected from population 1, and the same number of observations from population 2 so that, upon computation of the sample means, we can state that $(\mu_1 - \mu_2) = (\overline{X}_1 - \overline{X}_2) \pm 1.5$ with 99% confidence.

d. Sample size for $(\phi_1 - \phi_2)$

The proper sample size to estimate the difference between two population proportions is found by the following formula:

$$n = \frac{(Z_{\alpha/2})^2(\phi_1 q_1 + \phi_2 q_2)}{E^2}$$

In most practical applications we do not know ϕ. In those cases, we use the proportion of the pilot sample.

Example #4:
Ms. Y is a candidate for political office. She would like to find out how many women as well as men have to be sampled so that with 97% confidence, their difference in willingness to vote for her is equal to the difference in the proportions of the samples plus or minus 5%. If according to pilot sampling 47% (51%) of women (men) would vote for her, how many men and how many women should be sampled?

$$n = \frac{(2.17)^2[(0.47)(0.53) + (0.51)(0.49)]}{0.05^2} \approx 1,469$$

Thus, if 1,469 men and 1,469 women are sampled, then we can state that $(\phi_1 - \phi_2) = (p_1 - p_2) \pm 0.05$ with 97% confidence.

e.   Sample size for Health and Medical Applications

The minimum sample size for a given confidence level is

$$n = \frac{(Z_{\alpha/2}^2)(\text{Sensitivity})(1 - \text{Sensitivity})}{D^2}$$

where D represents the width of the confidence interval.

Example #5
Let us have a 95% confidence interval with a sensitivity = 0.9 and a width of 1%. Given this, α = 1-0.95=0.05 and $Z_{\alpha/2}$ = 1.96. Thus, the minimum sample size required is:

$$n = \frac{\left(1.96^2\right)\left(0.9\right)\left(0.1\right)}{0.01^2} = 3,457$$

For a width of 2%:

$$n = \frac{\left(1.96^2\right)\left(0.9\right)\left(0.1\right)}{0.02^2} \cong 865$$

Notice that the greater the width the less observations necessary and vice-versa.

3.9 GENERAL QUESTIONS AND ANSWERS

Question 1. The mean of a normal distribution X is 60. The SDM, with n = 9, has generated a standard error of 5. What is the probability that X is greater than 70 and less than 40?

Answer for #1:

$$P(X \geq 70) + P(X \leq 40) = P(Z \geq 0.67) + P(Z \leq -1.33)$$

$$= 0.2514 + 0.0918 = 0.3932$$

Question 2. The mean and variance of population X are 15 and 16 respectively.

a.   Compute $P(10 \leq X \leq 20)$.

b.   Compute $P(20 \leq X \leq 26)$.

c.   Compute $P(12 \leq X \leq 15)$.

Answers for #2:

a. $P(10 \leq X \leq 20) = P(0 \leq Z \leq 1.25) + P(-1.25 \leq Z \leq 0) = 2(0.3944) = 0.7888$.

b. $P(20 \leq X \leq 26) = P(0 \leq Z \leq 2.75) - P(0 \leq Z \leq 1.25) = 0.497 - 0.3944 = 0.1026$.

c. $P(12 \leq X \leq 15) = P(-0.75 \leq Z \leq 0) = 0.2734$.

Question 3. The upper 15% of grades in a Statistics exam will receive As and Bs. What is the lowest score to be designated a B? The scores are normally distributed with a mean of 70 and a standard deviation of 6.

Answer for #3:

$$\text{From } Z = \frac{(X - \mu)}{\sigma} \Rightarrow 1.04 = \frac{(X - 70)}{6} \Rightarrow X \approx 76 = \text{lowest B.}$$

Question 4. If the upper 10% of grades, in the above problem, are to receive As what is the lowest score to be designated an A?

Answer for #4:

$$\text{From } Z = \frac{(X - \mu)}{\sigma} \Rightarrow 2.33 = \frac{(X - 70)}{6} \Rightarrow X \approx 84 = \text{lowest A.}$$

Question 5. If the average is 80 and the standard deviation is 4, how would the results change in problems 4 and 5 above?

Answers for #5:

Lowest B ≈ 84, and lowest A ≈ 89.

Question 6. A population is normally distributed with a mean of 27 and a standard deviation of 5. If 36 observations are randomly selected from this population, what is the probability that the sample mean will be less than 25?

Answer for #6:

$$P(\overline{X} \leq 25) = P(Z \leq -2.4) = 0.4918,$$

$$\text{where } Z = \frac{(\overline{X} - \mu_{\overline{x}})}{\sigma_{\overline{x}}} = \frac{(\overline{X} - \mu)}{\frac{\sigma}{\sqrt{n}}} = \frac{(25 - 27)}{\frac{5}{\sqrt{36}}} = -2.4$$

Question 7. Suppose that 40% of all female students in the USA are expected to vote for candidate G. If $\mu_p = 0.4$ and $\sigma_p = 1.2$, what is the probability that a sample of $n = 500$ would yield a percentage greater than 50?

Answer for #7:

Because $0.4(500) = 200$ and $0.6(500) = 300$, we can use the Z distribution. Therefore,

$$P(p \geq 0.5) = P(Z \geq 0.08) = 0.5\text{-}0.0319 = 0.4681,$$

where $Z = \dfrac{(p - \mu_p)}{\sigma_p} = \dfrac{(0.5 - 0.4)}{1.2} = 0.08.$

Question 8. When df = 20, what is c equal to in $P(t \geq c) = 0.005$?

Answer for #8:

From the t table, c = 2.845.

Question 9. When df = 15, what are c_1 and c_2 in

$P(t \geq c_1) + P(t \leq c_2) = 0.05$?

Answer for #9:

From the t table, $c_1 = 2.131$ and $c_2 = 2.131$.

Question 10. Data for a sample, derived from a normal population without re-placement, is reported below. Compute the 99% confidence interval about the mean of the population.

Sample Data: 20, 25, 30, 12, 19, 17, 26, 33, 37, 22, 54, 19, 14, 10

Answer for #10:

n = 14, $\overline{X} = 24.143$, s = 11.6, df = n-1 = 13, and $t_{\alpha/2} = 3.012$.

Therefore,

$\mu = 24.143 \pm 9.34$ with 99% confidence.

Question 11. In a human sample of size n = 87, 26 tested negative. Compute the 96% confidence interval about the proportion of the population.

Answer for #11:

Since np and n(1-p) are both greater than 5 we can use the Z distribution. There-fore,

$\phi = 0.299 \pm 0.101$ with 96% confidence.

Question 12. A sample is summarized by the following information:
$n = 28$, and $s^2 = 49$. Compute the 99% confidence interval about the variance and standard deviation of the population.

Answer for #12:

$U = 0.005$, $L = 0.995$, $df = n-1 = 27$,

$\chi_U^2 = 49.645$, and $\chi_L^2 = 11.808$.

Therefore,

$P(26.65 \leq \sigma^2 \leq 112.04) = 0.99$, and

$P(5.16 \leq \sigma \leq 10.58) = 0.99$.

Question 13. Information for a sample is reported below. Compute the 95% and 99% confidence intervals about the mean and standard deviation of the popula-tion.

Sample information:

$n = 61$
$\overline{X} = 20$
$s^2 = 12$.

Answer for #13:

For μ:
$P(19.13 \leq \mu \leq 20.87) = 0.95$,
$P(18.86 \leq \mu \leq 21.14) = 0.99$.

For σ (use s):

$P(2.5 \le \sigma \le 5.1) = 0.95,$

$P(2.3 \le \sigma \le 5.8) = 0.95.$

Question 14. Some of the statistics generated from two different sampling procedures are reported below:

Procedure 1	Procedure 2
$n_1 = 64$	$n_2 = 81$
$\overline{X}_1 = 17$	$\overline{X}_2 = 13$
$s_1 = 5$	$s_2 = 4$

Use a 90%, and then a 99% confidence interval to estimate the difference in the means of the populations from which the samples were drawn.

Answers for #14:

$(\mu_1 - \mu_2) = 4 \pm 1.26$ with 90% confidence, and

$(\mu_1 - \mu_2) = 4 \pm 1.98$ with 99% confidence.

Question 15. Some of the statistics generated from two different samples, drawn from normal populations, are reported below:

Procedure 1	Procedure 2
$n_1 = 15$	$n_2 = 12$
$\overline{X}_1 = 4$	$\overline{X}_2 = 9$
$s_1 = 5$	$s_2 = 4.9$

Use a 99% confidence interval to estimate the difference in the means of the populations from which the samples were drawn.

Answer for #15:

Notice that $s_1 \approx s_2$. Therefore,

$(\mu_1 - \mu_2) = 3 \pm 5.35$ with 99% confidence.

Question 16. Some of the statistics generated from two different samples, drawn from normal populations, are reported below:

Procedure 1	Procedure 2
$n_1 = 25$	$n_2 = 17$
$\overline{X}_1 = 9$	$\overline{X}_2 = 4$
$s_1 = 3$	$s_2 = 12$

Use a 99% confidence interval to estimate the difference in the means of the populations from which the samples were drawn.

Answer for #16:

Notice that s_1 and s_2 are not approximately equal to each other. Therefore,

df $=16$ and $(\mu_1-\mu_2) = 5 \pm 8.68$ with 99% confidence.

Question 17. 86 out of 200 children sampled in country A carry a certain disease. 215 out of 520 children sampled in country B carry the same disease. Estimate, with 99% confidence, the difference in the population proportions from which the samples were drawn.

Answer for #17:

$(\phi_1-\phi_2) = 0.0165 \pm 0.1059$

Question 18. Assume that the acceptable tolerable error is 1, or $E = 1$. A pilot sampling has generated $s = 18$. What should the sample size be with 99% confidence? What if $E = 2$? What if $E = 0.5$?

Answers for #18:

When $E = 1: n = \left[\dfrac{(\sigma)(Z_{\alpha/2})}{E} \right]^2 = \left[\dfrac{(18)(2.575)}{1} \right]^2 = 2,148$; and

when $E = 2 : n = \left[\dfrac{(18)(2.575)}{2} \right]^2 = 537$; and when $E = 0.5 : n = \left[\dfrac{(18)(2.575)}{0.5} \right]^2 = 8{,}593$.

Question 19. A Pharmaceutical company would like to estimate within 5 percentage points, and with 99% confidence, the proportion of people cured by one of the company's popular medicines. How large should the sample size be?

Answer for #19:

$$n = \phi q \left(\frac{Z_{\alpha/2}}{E} \right)^2 = (0.5)(0.5) \left(\frac{2.575}{0.05} \right)^2 \approx 664 .$$

Question 20. A researcher believes that the difference in the means of populations 1 and 2 is minimal. Actually, she believes that the tolerable error is $\pm\, 0.02$. Pilot samples have generated $s_1 = 7$, and $s_2 = 9$. If she desires to construct a 99% confidence interval about the difference in the means of the populations, what should the sample sizes be? What if the tolerable error is $\pm\, 0.08$?

Answers for #20:

$$n = \frac{(Z_{\alpha/2})^2 (\sigma_1^2 + \sigma_2^2)}{E^2}$$

With $E = 0.02 : n = \dfrac{(2.575)^2 (7^2 + 9^2)}{0.02^2} = 2{,}155{,}075$; and

With $E = 0.08 : n = \dfrac{(2.575)^2 (7^2 + 9^2)}{0.08^2} = 134{,}692$.

Question 21. In pilot sampling, 52% boys and only 22% girls carry a certain disease. How many boys and how many girls have to be sampled so that with 95% confidence the difference in the proportions of the populations is equal to the difference in the proportions of the samples plus or minus 2%? What if $E = 0.04$?

Answers for #21:

$$n = \frac{(Z_{\alpha/2})^2 (\phi_1 q_1 + \phi_2 q_2)}{E^2}$$

With $E = 0.02$: $n = \dfrac{(1.96)^2 [(0.52)(0.48) + (0.22)(0.78)]}{0.02^2} \approx 4{,}045$; and

with $E = 0.04$: $n = \dfrac{(1.96)^2 [(0.52)(0.48) + (0.22)(0.78)]}{0.04^2} \approx 1{,}011$.

3.10 HEALTH AND MEDICAL QUESTIONS AND ANSWERS

Question 1. The duration of cigarette smoking has been linked to many diseases, including lung cancer and various forms of heart disease. Assume that, based on previous research among men ages 30 to 34 who have ever smoked, the mean number of years they should smoke is 12.8 with a standard deviation of 5.1 years. For women in this age group, the mean number of years they should smoke is 9.3 with a standard deviation of 3.2.

a. Assuming that the duration of smoking is normally distributed, what proportion of men in this age group have smoked for more than 20 years?

b. Answer part **a** for women.

Answer for #1:

a. $Z = \dfrac{(X - \mu)}{\sigma} = \dfrac{(20 - 12.8)}{5.1} = 1.41$

$P(Z \geq 20) = 0.0793$

b. $Z = \dfrac{(20 - 9.3)}{3.2} = 3.34$

$P(Z \geq 20) = 0.0004$

Question 2. Serum cholesterol is an important risk factor for coronary disease. We can show that serum cholesterol is approximately normally distributed with mean 219 mg%/mL and standard deviation of 50 mg%/mL.

a. If the clinically desirable range for cholesterol is less than 200 mg%/mL, then what proportion of people have clinically desirable levels of cholesterol?

b. Some researchers feel that only cholesterol levels of over 250 mg%/mL indicate high enough risk for heart disease to warrant treatment. What proportion of the population does this group represent?

c. What proportion of the general population has borderline high cholesterol levels? Specifically, levels greater than 200, but less than 250 mg%/mL.

Answer for #2:

a. $Z = \dfrac{(X - \mu)}{\sigma} = \dfrac{(200 - 219)}{50} = -0.38$

P(Z ≤ 200) = 0.3520

b. $Z = \dfrac{(250 - 219)}{50} = 0.62$

P(Z ≥ 250) = 0.2676

c. P(200 ≤ Z ≤ 250) = 1- (0.3520 + 0.2676) = 0.3804

Question 3. Suppose we wish to estimate the concentration (µg/mL) of a specific dose of ampicillin in the urine after various periods of time. We recruit 125 volunteers and find that they have a mean concentration of 7.0 µg/mL with a standard deviation of 2.0 µg/mL. Assume that the underlying population distribution of concentrations is normally distributed

a. Find a 95% confidence interval for the population mean concentration.

b. How large a sample would be needed to ensure that the length of the confidence interval in part a is 0.5 µg/mL if we assume that the sample standard deviation remains at 2.0 µg/mL?

Answer for #3:
Using the basic formula of the confidence interval for the population mean:

$$P\left[\overline{X} - Z_{\alpha/2}\left(\frac{\sigma}{\sqrt{n}}\right) \le \mu \le \overline{X} + Z_{\alpha/2}\left(\frac{\sigma}{\sqrt{n}}\right)\right] = 1 - \alpha$$

a. $P\left[7.0 - 1.96\left(\dfrac{2.0}{\sqrt{125}}\right) \le \mu \le 7.0 + 1.96\left(\dfrac{2.0}{\sqrt{125}}\right)\right] = 0.95$

⇒ P[6.65 ≤ µ ≤ 7.35] = 0.95

b. $P\left[7.0-1.96\left(\dfrac{2.0}{\sqrt{n}}\right) \le \mu \le 7.0+1.96\left(\dfrac{2.0}{\sqrt{n}}\right)\right] = 0.95$

Thus, solving for n, we find that:

$1.96\left(\dfrac{2.0}{\sqrt{n}}\right) = 0.5 \Rightarrow \sqrt{n} = 7.84 \Rightarrow n \approx 61.5$

Question 4. Let us assume that a sample of 100 patients used for a particular study yields a sample sensitivity of 0.767. Construct a 95% confidence interval for the sensitivity of the population.

Answer for #4:
The sample standard deviation is :

$s = \sqrt{\dfrac{(0.767)(0.233)}{100}} = 0.042$

The 95% confidence interval for the sensitivity of the population is

$P[0.767-1.96(0.042) \le \text{Sensitivity}_{\text{Population}} \le 0.767+1.96(0.042)] = 0.95$

$P[\, 0.68468 \le \text{Sensitivity}_{\text{Population}} \le 0.84932 \,] = 0.95$

Chapter 4

HYPOTHESIS TESTING

"The only relevant test of the validity of a hypothesis is the comparison of its predictions with experience"

-Milton Friedman

An informed (or educated) guess about a population parameter, tentatively advanced as possibly true, is called *Statistical Hypothesis*. Utilizing a *Hypothesis Testing Procedure*, the decision maker may accept or reject such a hypothesis.

4.1 Hypothesis Testing Procedure

Step 1: Formulate the null and the alternative hypotheses. The *null hypothesis* (H_0) states something about a population parameter: it is an informed guess about a population parameter which is held true until sufficient statistical evidence shows otherwise. The *alternative hypothesis* (H_1) states something different: it is the informed guess of all contingencies not covered by the null hypothesis.

If ω = population parameter, a hypothesis test may be stated in one of the following three forms:

Two-tail test:	H_0: $\omega = \omega_0$
	H_1: $\omega \neq \omega_0$
Left-tail test:	H_0: $\omega \geq \omega_0$
	H_1: $\omega < \omega_0$
Right-tail test:	H_0: $\omega \leq \omega_0$
	H_1: $\omega > \omega_0$

Step 2: Select the *significance level*, α, at which the test is to be conducted, where α = probability of rejecting the H_0 when in fact the H_0 is true, and $(1-\alpha)$ = confidence level of the hypothesis test. (More on α follows in the next section).

Step 3: From the Z, or t, or χ^2 table determine the *rejection region* for H_0. The rejection region consists of the values of the test statistic for which the null hypo-

thesis is rejected. In general, the rejection region is the region that corresponds
to the α area.

Two-tail test:

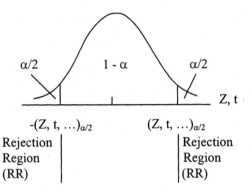

Left-tail test:

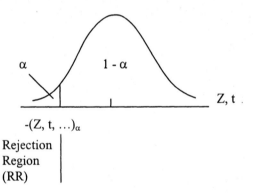

Right-tail test:

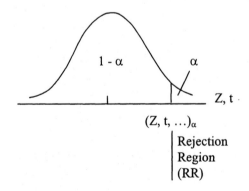

Step 4: Draw a random sample, and calculate the value of the test statistic.
The test statistic is the random variable whose value determines the conclusion.
If the value of the test statistic falls in the rejection region, the H_0 is rejected and
the H_1 is accepted; if it does not fall in the rejection region, insufficient evidence
exists to support the alternative hypothesis and therefore the H_0 is not rejected.

When $n > 30$ and $n < 0.05N$, the test statistic for $\omega = \mu$ is:

$$Z_{st} = \frac{\left(\overline{X} - \mu_0\right)}{\left(\dfrac{\sigma}{\sqrt{n}}\right)}$$

where nearly always the population standard deviation must be approximated by
the sample standard deviation s.

When $n < 30$, and the parent population is normal, the test statistic, with
$df = n-1$, for $\omega = \mu$ is:

$$t_{st} = \frac{\left(\overline{X} - \mu_0\right)}{\left(\dfrac{s}{\sqrt{n}}\right)}.$$

When $n\phi \geq 5$ and $n(1-\phi) \geq 5$, the test statistic for $\omega = \phi$ is:

$$Z_{st} = \frac{\left(p - \phi_0\right)}{\sqrt{\left[\dfrac{p(1-p)}{n}\right]}}$$

When $\omega = \sigma^2$, and the parent population is normal, the test statistic, with
$df = n - 1$, is:

$$\chi^2_{st} = \frac{\left[(n-1)s^2\right]}{\sigma_0^2}$$

Important Note: Statistical software packages provide the so called *p-value* for a
hypothesis test. The p-value is the smallest possible choice of α for which H_0 can
be rejected. The p-value is also called the *observed significance level.* When the
p-value $< \alpha$, the H_0 is rejected.

Example #1:

A sample of n = 47 has been drawn from a population with $\overline{X}$ = 16 and s = 9. Test at the 5% level of significance whether or not the mean of the population is 18.

This is a two-tail test with μ_0 = 18, α = 0.05, 1-α = 0.95, and $Z_{\alpha/2}$ = ±1.96.

$$H_0: \mu = 18$$
$$H_1: \mu \neq 18$$

$$Z_{st} \frac{(\overline{X} - \mu_0)}{\left(\frac{\sigma}{\sqrt{n}}\right)} = \frac{(16 - 18)}{\left(\frac{9}{\sqrt{47}}\right)} = -1.52 > Z_{\alpha/2} = -1.96.$$

Thus, the H_0 may not be rejected.

Example #2:

You are about to start your second year in college. Your grade point average, based on the 10 courses you have taken so far, is 2.7, with a standard deviation of 0.5. If you expect your grades to be normally distributed, can you claim that your real GPA is at least 3.0? Test your claim at α = 0.01.

This is a left-tail test with μ_0 = 3, and we can use the t distribution with df = 9.

$$H_0: \mu \geq 3$$
$$H_1: \mu < 3$$

At α = 0.01, t_α = -2.821, and

$$t_{st} = \frac{(\overline{X} - \mu_0)}{\left(\frac{s}{\sqrt{n}}\right)} = \frac{(2.7 - 3)}{\left(\frac{0.5}{\sqrt{10}}\right)} = -1.897 > t\alpha = -2.821.$$

Thus, you fail to reject the H_0 and therefore, your claim may be valid.

Example #3:

A "counting devise" has been installed in a shopping mall which records the number of daily visitors. The manager of the mall suspects that the counting devise overstates the number of visitors. A sample of 35 randomly selected days generated a mean of 3,000 people and a standard deviation of 150. Can she claim, with 99% confidence, that the daily average number of visitors is no more than 2,500?

This is a right-tail test with $\mu_0 = 2,500$, $\alpha = 0.01$, $1-\alpha = 0.99$, and $Z_\alpha = \pm 2.575$.

$$H_0: \mu \leq 2,500$$
$$H_1: \mu > 2,500$$

$$Z_{st} = \frac{\left(\overline{X} - \mu_0\right)}{\left(\dfrac{\sigma}{\sqrt{n}}\right)} = \frac{\left(3,000 - 2,500\right)}{\left(\dfrac{150}{\sqrt{35}}\right)} = 19.72 > Z_\alpha = 2.575.$$

Thus, the H_0 is rejected in favor of the alternative, and so is the manager's claim that the average number of visitors is no more than 2,500.

Example #4:
Consider the previous example. Can the manager claim, with the same confidence, that the daily average number of visitors is no more than 2,950?

In this case,

$$H_0: \mu \leq 2,950$$
$$H_1: \mu > 2,950$$

$$Z_{st} = \frac{\left(\overline{X} - \mu_0\right)}{\left(\dfrac{\sigma}{\sqrt{n}}\right)} = \frac{\left(3,000 - 2,950\right)}{\left(\dfrac{150}{\sqrt{35}}\right)} = 1.97 < Z_\alpha = 2.575.$$

Thus, the H_0 may not be rejected and so is the manager's claim.

Example #5:
The proportion of interest in a sample of 75 observations is 0.8. Can we claim, with 95% confidence, that the population's proportion of interest is different than 0.7?

This is a two-tail test with $\phi_0 = 0.7$, $\alpha = 0.05$, $1-\alpha = 0.95$, and $Z_{\alpha/2} = \pm 1.96$.

$$H_0: \phi = 0.7$$
$$H_1: \phi \neq 0.7$$

$$Z_{st} = \frac{(p - \phi_0)}{\sqrt{\left[\dfrac{p(1-p)}{n}\right]}} = \frac{(0.8 - 0.7)}{\sqrt{\left[\dfrac{0.8(1-0.8)}{75}\right]}} = 2.17 > Z_{\alpha/2} = 1.96.$$

Thus, the H_0 is rejected and our claim, that the population's proportion of interest is different than 0.7, is accepted.

Example #6:
In the above example, can we claim, with the same confidence, that the population's proportion of interest is less than 0.7?

In this case,

$$H_0: \phi \geq 0.7$$
$$H_1: \phi < 0.7$$

$$Z_{st} = \frac{(p - \phi_0)}{\sqrt{\left[\dfrac{p(1-p)}{n}\right]}} = \frac{(0.8 - 0.7)}{\sqrt{\left[\dfrac{0.8(1-0.8)}{75}\right]}} = 2.17 > Z_{\alpha} = -1.645.$$

Thus, the H_0 may not be rejected and our claim may not be accepted.

Example #7:
In the above example, can we claim, with the same confidence, that the population's proportion of interest is more than 0.7?

In this case,

H_0: $\phi \leq 0.7$
H_1: $\phi > 0.7$

$$Z_{st} = \frac{(p - \phi_0)}{\sqrt{\left[\dfrac{p(1-p)}{n}\right]}} = \frac{(0.8 - 0.7)}{\sqrt{\left[\dfrac{0.8(1-0.8)}{75}\right]}} = 2.17 > Z_\alpha = 1.645.$$

Thus, the H_0 is rejected and our claim is accepted.

Example #8:
A sample of n = 95, from a normal population, has a variance of 2,996.58. Can we claim, with 95% confidence, that the variance of the population is about 3,600?

This is a two-tail test with $\alpha = 0.05$, df = 94, and from the χ^2 distribution:

$\chi_U^2 = 118.14$, $\chi_L^2 = 65.65$.

H_0: $\sigma^2 = 3,600$
H_1: $\sigma^2 \neq 3,600$

$$\chi^2_{st} = \frac{\left[(n-1)s^2\right]}{\sigma_0^2} = \frac{(94)(2,996.58)}{3,600} = 78.24$$

which is greater than χ_L^2 and less than χ_U^2.

Therefore, the H_0 may not be rejected and so is our claim.

Example #9:
In the above example, can we claim, with the same confidence, that the variance of the population is greater than 3,700?

In this case,

$$H_0: \sigma^2 \leq 3,700$$
$$H_1: \sigma^2 > 3,700$$

Since $\chi^2_{st} = 76.13 < \chi^2_{\alpha} = 113.14$, the H_0 may not be rejected and therefore our claim may not be accepted.

4.2 Two Types of Errors

Whenever incomplete information such as a sample is used to make an inference about a population, there is a risk of making a mistake. In a problem involving a test of hypothesis, there are two types of erroneous conclusions that could be made.

Type I error occurs when we reject a true null hypothesis, or
P(Type I error) = P(Reject H_0 when H_0 is true) = α.

Type II error occurs when we accept a false null hypothesis, or
P(Type II error) = P(Accept H_0 when H_0 is false) = β.

β is computed as follows:

For a left-tail test:

$$\beta = P\left[Z \geq \frac{(\mu_o - \mu_r)}{\left(\dfrac{s}{\sqrt{n}}\right)} - |Z_\alpha| \right]$$

For a right-tail test:

$$\beta = P\left[Z \leq \frac{(\mu_o - \mu_r)}{\left(\dfrac{s}{\sqrt{n}}\right)} + |Z_\alpha| \right]$$

For a two-tail test:

$$\beta = P\left\{\left[\frac{(\mu_o - \mu_r)}{\left(\frac{s}{\sqrt{n}}\right)} - \left|Z_{\alpha/2}\right|\right] \leq Z \leq \left[\frac{(\mu_o - \mu_r)}{\left(\frac{s}{\sqrt{n}}\right)} + \left|Z_{\alpha/2}\right|\right]\right\}$$

where μ_r is a number different than μ_0. Let the difference between μ_0 and μ_r be Δ, so that $\Delta = \mu_0 - \mu_r$.

The *Power* (P_r) of a statistical hypothesis test is the probability of rejecting the null hypothesis when the null hypothesis is false.

$$\mathbf{P_r = 1 - \beta}$$

The size of P_r depends on the following:

(i)   The greater the Δ differential, the greater the power.
(ii) The smaller the standard deviation, the greater the power.
(iii) The larger the sample size, the greater the power.
(iv) The smaller the α, the smaller the power.

The objective of the researcher is to maximize P_r and minimize sampling costs.

Important Notes:
a.   In general, the hypothesis-testing procedure outlined so far can lead to one of four probabilistic results:

(i)   P(Reject H_0 when H_0 is true) = α = Type I error
(ii) P(Accept H_0 when H_0 is true) = 1- α = Confidence Level
(iii) P(Accept H_0 when H_0 is false) = β = Type II error
(iv) P(Reject H_0 when H_0 is false) = 1- β = Power

(α is also called "*significance level*," whereas β is also called "*error of acceptance*")

b.   The researcher should not be concerned with the computation of β when the H_0 is rejected. Rejection of the H_0 is considered acceptable when $\alpha \leq 0.05$. Inability to reject the H_0 implies β or error of acceptance. In this case only a low β (high Power) would make the conclusion credible. The H_0 is "accepted" when $\beta \leq 0.05$.

Example #1:
A sample of size n = 95 has a mean of $\overline{X}$ = 67.632, and a standard deviation of
s = 54.741.

a.   Let μ_0 = 65. Show, with 90% confidence, that the H_0 of the right-tail test
may not be rejected.

b.   What if μ_τ = 66? Compute the Power of the test.

c.   Next compute the Power of the test when μ_τ = 66 and α = 0.05.

d.   Next compute the Power of the test when μ_τ = 67 and α = 0.05.

e.   Let μ_0 = 65. Show, with 90% confidence, that the H_0 of the two-tail test may
not be rejected and then compute the Power of the test when μ_τ = 66.

Answers for Example #1:

a.   From the Z distribution Z_α = 1.28 which is greater than Z_{st} = 0.47.

      Therefore the H_0 may not be rejected.

b.   Because this is a right-tail test,

$$\beta = P\left[Z \geq \frac{(\mu_o - \mu_\tau)}{\left(\dfrac{s}{\sqrt{n}}\right)} + |Z_\alpha| \right]$$

$$\Rightarrow P\left[Z \leq \frac{(65 - 66)}{\left(\dfrac{54.741}{\sqrt{95}}\right)} + 1.28 \right] = P(Z \leq 1.102) =$$

P(- ∞ ≤ Z ≤ 0)+ P(0 ≤ Z ≤ 1.102) = 0.3643+0.5 = 0.8643.

Therefore, Power = 1-β = 1-0.8643 = 0.1357.

c.   β = 0.9279, and Power = 0.0721.

d. $\beta = 0.8997$, and Power $= 0.1003$.

e. From the Z distribution, $Z_{\alpha/2} = \pm 1.64$ which, in absolute value, is greater than $Z_{st} = 0.47$. Therefore the H_0 may not be rejected.

$$\Rightarrow \beta = P\left\{ \left[\frac{(\mu_0 - \mu_\tau)}{\left(\dfrac{s}{\sqrt{n}}\right)} - \left|Z_{\alpha/2}\right| \right] \leq Z \leq \left[\frac{(\mu_0 - \mu_\tau)}{\left(\dfrac{s}{\sqrt{n}}\right)} + \left|Z_{\alpha/2}\right| \right] \right\}$$

$= P(-1.82 \leq Z \leq 1.46) = 0.8935$. Therefore, Power $= 0.1065$.

Example #2:
The average of a sample of 290 observations is $\overline{X} = 56$; its standard deviation is $s = 2$.

a. Test, at the 5% level, whether $\mu < 56$.

b. What is the Power of the test when: $\mu_\tau = 50$, $\mu_\tau = 55.5$, and $\mu_\tau = 56.1$?

Answers for Example #2:

a. $Z_\alpha = -1.64$ is less than $Z_{st} = 0$. Therefore, the H_0 may not be rejected.

b. $\beta = P\left[Z \geq \dfrac{(\mu_0 - \mu_\tau)}{\left(\dfrac{s}{\sqrt{n}}\right)} - \left|Z_\alpha\right| \right]$.

Thus,

at $\mu_\tau = 50$, $\beta = P(Z \geq 49.45) = 0$ and Power $= 1$;

at $\mu_\tau = 55.5$, $\beta = P(Z \geq 2.62) = 0.0044$ and Power $= 0.9956$;

at $\mu_\tau = 56.1$, $\beta = P(Z \geq -2.49) = 0.9936$ and Power $= 0.0064$.

4.3 Tests of Hypotheses for *Two* Population Parameters

If ω_1 and ω_2 are two population parameters, a hypothesis test may be stated one of the following three forms:

Two-tail test: H_0: $\omega_1 = \omega_2$ or H_0: $\omega_1 - \omega_2 = \omega_0$
$\qquad\qquad\qquad$ H_1: $\omega_1 \neq \omega_2$ $\qquad$ H_1: $\omega_1 - \omega_2 \neq \omega_0$

Left-tail test: H_0: $\omega_1 \geq \omega_2$
$\qquad\qquad\qquad$ H_1: $\omega_1 < \omega_2$

Right-tail test: H_0: $\omega_1 \leq \omega_2$
$\qquad\qquad\qquad$ H_1: $\omega_1 > \omega_2$

a.   The Z statistic for $(\mu_1-\mu_2)$ with $30 \leq n_1 < 0.05N_1$, and $30 \leq n_2 < 0.05N_2$ is:

$$Z_{st} = \frac{\left[(\overline{X}_1 - \overline{X}_2) - \mu_0\right]}{\sqrt{\left[\left(\dfrac{s_1^2}{n_1}\right) + \left(\dfrac{s_2^2}{n_2}\right)\right]}}$$

If the population variances are known, regardless of sample size:

$$Z_{st} = \frac{\left[(\overline{X}_1 - \overline{X}_2) - \mu_0\right]}{\sqrt{\left[\left(\dfrac{\sigma_1^2}{n_1}\right) + \left(\dfrac{\sigma_2^2}{n_2}\right)\right]}}$$

b.   With sample sizes of $n_1 < 30$ or $n_2 < 30$, drawn from normal populations with equal or approximately equal sigmas, and df $= n_1 + n_2 - 2$, the t statistic for $(\mu_1-\mu_2)$ is:

$$t_{st} = \frac{\left[(\overline{X}_1 - \overline{X}_2) - \mu_0\right]}{\sqrt{\left\{s_p^2\left[\left(\dfrac{1}{n_1}\right) + \left(\dfrac{1}{n_2}\right)\right]\right\}}}$$

where s_p^2 is computed as in section 3.6.

c. With sample sizes of $n_1 < 30$ or $n_2 < 30$, drawn from normal populations, the t statistic for $(\mu_1 - \mu_2)$, with df computed as in Chapter 3 is:

$$t_{st} = \frac{\left[(\overline{X}_1 - \overline{X}_2) - \mu_0\right]}{\sqrt{\left[\left(\dfrac{s_1^2}{n_1}\right) + \left(\dfrac{s_2^2}{n_2}\right)\right]}}$$

d. The Z statistic for two population proportions $(\phi_1 - \phi_2)$, with sample sizes of $30 \leq n_1 < 0.05N_1$, and $30 \leq n_2 < 0.05N_2$ is:

$$Z_{st} = \frac{\left[(p_1 - p_2) - \phi_0\right]}{\sqrt{\left[\left(\dfrac{p_1(1 - p_1)}{n_1}\right) + \left(\dfrac{p_2(1 - p_2)}{n_2}\right)\right]}}$$

When $\phi_0 = 0$,

$$Z_{st} = \frac{(p_1 - p_2)}{\sqrt{\left[\left(\dfrac{p_1(1 - p_1)}{n_1}\right) + \left(\dfrac{p_2(1 - p_2)}{n_2}\right)\right]}} ,$$

$$p = \frac{(x_1 + x_2)}{(n_1 + n_2)} , \text{ where } x_i \ (i = 1,2) \text{ are the numbers in the first and}$$

second sample with the characteristic of interest.

e. The F statistic, for two population variances with $df_n = n_1 - 1$, and $df_d = n_2 - 1$ (where df_n = degrees of freedom for numerator, and df_d = degrees of freedom for denominator), is:

$$F_{st} = \left(\frac{S_1^2}{S_2^2}\right), \text{ where } S_1^2 \text{ is the larger sample variance.}$$

The F Distribution table is used to find areas under the F curve (see Appendix III). For instance, the 95% percentile for the F distribution having $df_n = 7$ and $df_d = 15$ is $F_\alpha = 2.71$. The 99% percentile for the same distribution with the same degrees of freedom is $F_\alpha = 4.14$.

Example #1:
Two samples of sizes $n_1 = 12$, and $n_2 = 16$ have been derived from two populations with known variances: $\sigma^2_1 = 36$ and $\sigma^2_2 = 64$. The sample means are, respectively, 40 and 34. Test, at $\alpha = 0.05$ whether or not the two samples come
from the same population.

H_0: $\mu_1 - \mu_2 = 0$
H_1: $\mu_1 - \mu_2 \neq 0$

$Z_{\alpha/2} = \pm 1.96$ and $Z_{st} = 2.27$.

Therefore, because 2.27 falls to the right of 1.96 the H_0 is rejected. This
implies that the samples do not come from the same population.

Example #2:
Two samples of sizes $n_1 = 50$, and $n_2 = 60$ have been derived from two populations with unknown variances. The sample means are, respectively, 11 and 9.
The sample standard deviations are, respectively, 1 and 0.5. Test, at $\alpha = 0.01$,
your claim that the difference between the populations is greater than 0.5.

H_0: $\mu_1 - \mu_2 \leq 0.5$
H_1: $\mu_1 - \mu_2 > 0.5$

$Z_\alpha = 2.33$ and $Z_{st} = 9.649$.

Therefore, because 9.649 falls to the right of 2.33 the H_0 is rejected in
favor of your claim.

Example #3:
Sample information from two normally distributed populations, with approximately equal variances, is given below:

Sample 1	Sample 2
$n_1 = 10$	$n_2 = 12$
$s_1 = 5$	$s_2 = 4$
$\overline{X}_1 = 100$	$\overline{X}_2 = 98$

Test, with 99% confidence, your claim that the samples do not come from the
same populations.

H_0: $\mu_1 - \mu_2 = 0$
H_1: $\mu_1 - \mu_2 \neq 0$

df = 10+12-2 = 20, $t_{\alpha/2} = \pm 2.845$ and $\mu_0 = 0$

$$\Rightarrow t_{st} = \frac{\left[(\overline{X}_1 - \overline{X}_2) - \mu_0\right]}{\left\{s_p^2\left[\left(\frac{1}{n_1}\right) + \left(\frac{1}{n_2}\right)\right]\right\}} = 1.04$$

Therefore, because 1.04 falls between $\pm$ 2.845 the H_0 may not be rejected, and your claim may not be accepted.

Example #4:
54 out of 225 urban families and 52 out of 175 rural families state that they do most of their grocery shopping at chain stores. Do the two groups differ? (Let $\alpha = 0.05$)

H_0: $\phi_1 - \phi_2 = 0$
H_1: $\phi_1 - \phi_2 \neq 0$

$\phi_0 = 0$, $Z_{\alpha/2} = \pm 1.96$, $p = \dfrac{(x_1 + x_2)}{(n_1 + n_2)} = \dfrac{(54 + 52)}{(225 + 175)} = 0.265$,

and $Z_{st} = -7.76$.

Because -7.76 is less than -1.96, the H_0 is rejected. Thus, the groups may not differ.

Example #5:
Consider the following survey results for areas 1 and 2:

Sample	Number of Households with Two or More Cars
$n_1 = 150$	113
$n_2 = 160$	104

Test, with 95% confidence, your claim that the proportion of area 1 exceeds by more than 0.05 the proportion of area 2.

H_0: $\phi_1 - \phi_2 \leq 0.05$
H_1: $\phi_1 - \phi_2 > 0.05$

$\phi_0 = 0.05$, $Z_\alpha = 1.645$, $Z_{st} = 1$.

Because 1 is less than 1.645 the H_0 may not be rejected, and therefore your claim may not be accepted.

Example #6:
The variances of two samples of sizes $n_1 = 22$ and $n_2 = 25$, derived from two normally distributed populations, are, respectively, 70.3 and 225.3. Can we claim, with 95% confidence, that the population variances are identical?

The hypotheses are:

$$H_0: \sigma^2_1 = \sigma^2_2$$
$$H_1: \sigma^2_1 \neq \sigma^2_2$$

$df_n = 21$, $df_d = 24$, $F_{\alpha/2} = \pm 2.33$ and $F_{st} = 225.3/70.3 = 3.21$.

Because 3.21 falls to the right of 2.33 the H_0 is rejected and so is our claim.

4.4 Tests of Hypotheses for Health and Medical Applications

Like the general hypothesis testing procedure outlined in section 4.1, we can test hypotheses for population sensitivity. In section 1.3, we defined sensitivity and specificity and illustrated how they can be computed from a matrix composed of number of observations from the sample that fall into each category/cell. These matrices can be represented generally as

	Variable 2 – Yes	Variable 2 - No	Total
Variable 1 –Yes	a	b	a+b
Variable 1 – No	c	d	c+d
Total	a+c	b+d	n

where: $n = (a+b+c+d)$.

We wish to use the matrix of observations to test the following hypotheses:

$$H_0: p_1 = p_2$$
$$H_1: p_1 \neq p_2$$

where p_1 is the probability of a "yes" with respect to variable 1 and p_2 is the probability of a "no" with respect to variable 1, *given variable 2 is "yes."* In other words, we wish to test the null hypothesis that these probabilities do not differ given variable 2 has a "yes" value. Thus, if the null hypothesis is true, we do not expect any relationship between variable 1 and variable 2.

The test statistic in these cases is distributed χ^2 and is given by

$$\chi^2_{st} = \frac{n\left[|ad - bc| - \left(\dfrac{n}{2}\right)\right]^2}{(a + c)(b + d)(a + b)(c + d)}$$

with df = (v-1)(c-1)

where v = number of rows in matrix and c = number of columns in matrix.

Here, with a v = 2 and c = 2 $\Rightarrow$ df = (2-1)(2-1) = 1.

Example #1:
The following table shows the results of a study in which researchers examined a child's IQ and the presence of a specific gene in the child.

	Gene Present	Gene Not Present	Total
High IQ	25	15	40
Normal IQ	10	20	30
Total	35	35	70

H_0: $p_1 = p_2$
H_1: $p_1 \neq p_2$

where p_1 is the probability of a having a high IQ and p_2 is the probability of having a normal IQ, given the gene is present.

If the null hypothesis is true, we do not expect the probability of having a high IQ and normal IQ to differ if the gene is present.

$$\chi^2_{st} = \frac{70\left[|(25)(20) - (15)(10)| - \left(\dfrac{70}{2}\right)\right]^2}{(35)(35)(40)(30)} = 4.725$$

Testing the null hypothesis with an $\alpha = 0.05$ and df $= 1$, the value from the χ^2 table is 3.84. Since $4.725 > 3.84 \Rightarrow$ we reject the null hypothesis. Thus, we believe there is a statistically significant relationship between having a high IQ and the gene being present.

4.5 Sample Size for Specific Level of Significance and Power

From the previous section, it is implied that

$$Z_\alpha = \frac{\left(\overline{X}_1 - \mu_0\right)}{\left(\dfrac{\sigma}{\sqrt{n}}\right)} \text{ and } Z_\beta = Z_\alpha = \frac{\left(\overline{X}_1 - \mu_\tau\right)}{\left(\dfrac{\sigma}{\sqrt{n}}\right)}$$

Solving for $\overline{X}_1$ we have:

$$\overline{X}_1 = \mu_0 + Z_\alpha \left(\frac{\sigma}{\sqrt{n}}\right) \text{ and } \overline{X}_1 = \mu_\tau + Z_\beta \left(\frac{\sigma}{\sqrt{n}}\right)$$

Setting the right sides equal to each other and solving for n we get:

$$n = \frac{\left[\left(Z_\alpha + Z_\beta\right)^2 \sigma^2\right]}{\Delta^2}, \text{ where } \Delta = |\mu_0 - \mu_\tau|.$$

Example #1:
Suppose we wish to test, at the $\alpha = 0.05$ level of significance, the hypothesis H_0: $\mu \leq 100$ versus H_1: $\mu > 100$

Let the population be normal mean μ and standard deviation $\sigma = 15$. Find the minimum sample size required if the power of the test is to be 0.95 when $\mu_\tau = 110$.

Notice that since $\alpha = \beta = 0.05$, $Z_\alpha = Z_\beta = 1.645$. Therefore,

$$n = \frac{\left[\left(Z_\alpha + Z_\beta\right)^2 \sigma^2\right]}{\Delta^2} = \frac{\left[(1.645 + 1.645)^2 15^2\right]}{10^2} = 24.35 \text{ or } 25.$$

For experiments about μ, with σ known or estimated from a pilot sample, sample size may be computed as follows:

Two-tail test: $\quad n = \dfrac{\left[\left(Z_{\alpha/2} + Z_{\beta/2}\right)^2 \sigma^2\right]}{\Delta^2}$

Right-tail test: $\quad n = \dfrac{\left[\left(Z_{\alpha} + Z_{\beta}\right)^2 \sigma^2\right]}{\Delta^2}$

Left-tail test: $\quad n = \dfrac{\left[\left(Z_{\alpha} + Z_{\beta}\right)^2 \sigma^2\right]}{\Delta^2}$

For experiments about $\mu_1 - \mu_2$, with σ's known or estimated from pilot samples, sample size, assuming that $n = n_1 = n_2$, may be computed as follows:

Two-tail test: $\quad n = \dfrac{\left[\left(Z_{\alpha/2} + Z_{\beta/2}\right)^2 \left(\sigma_1^2 + \sigma_2^2\right)^2\right]}{\Delta_1^2}$,

where $\Delta_1 = |\mu_1 - \mu_2|$

Right-tail test: $\quad n = \dfrac{\left[\left(Z_{\alpha} + Z_{\beta}\right)^2 \left(\sigma_1^2 + \sigma_2^2\right)^2\right]}{\Delta_1^2}$

Left-tail test: $\quad n = \dfrac{\left[\left(Z_{\alpha} + Z_{\beta}\right)^2 \left(\sigma_1^2 + \sigma_2^2\right)^2\right]}{\Delta_1^2}$

4.6 Diagnostic/Lab Tests versus Hypothesis Tests

	Diagnostic/Lab Test	*Hypothesis Test*
1. Scope	A test is performed on a single individual and yields a diagnosis: positive or negative.	A test is performed from a single experiment and yields a p-value: significant or insignificant.
2. Threshold	The pre-specified value that divides "normal" and "abnormal" is based on the relative opportunity cost of false-positive and false-negative diagnoses.	The pre-specified level of significance (α) that divides "significant" and "insignificant" is based on the relative opportunity cost of Type I and Type II errors.
3. Result	A measurement that can be compared to a pre-specified threshold and classified as "normal" or "abnormal."	A p-value that can be compared to a pre-specified level of significance (α) and classified as "significant" or "insignificant."
4. Errors	False-positives and false-negatives	Type I and Type II
5. Accuracy	Sensitivity and specificity	Level of significance (α) and power of test (β).

Source: Modified from Motulsky (1995).

4.7 GENERAL QUESTIONS AND ANSWERS

Sample data from populations X and Y, and their statistics, are reported below:

X	Y		X	Y
23.000	16.000	N OF CASES	32	32
12.000	18.000	MINIMUM	10.000	9.000
49.000	23.000	MAXIMUM	87.000	76.000
47.000	17.000	RANGE	77.000	67.000
10.000	24.000	MEAN	40.344	29.094
13.000	56.000	VARIANCE	412.168	282.926
36.000	19.000	STANDARD DEVIATION	20.302	16.820
78.000	24.000	STD. ERROR	3.589	2.973
11.000	23.000	SKEWNESS (G1)	0.558	1.158
56.000	12.000	KURTOSIS (G2)	-0.179	0.500
87.000	13.000	SUM	1,291.000	931.000
34.000	17.000	C.V.	0.503	0.578
37.000	18.000	MEDIAN	38.000	23.000
45.000	9.000			
39.000	44.000			
56.000	24.000			
23.000	46.000			
51.000	23.000			
54.000	13.000			
23.000	43.000			
45.000	25.000			
44.000	76.000			
23.000	23.000			
34.000	45.000			
66.000	23.000			
23.000	43.000			
24.000	65.000			
84.000	23.000			
34.000	56.000			
29.000	34.000			
45.000	24.000			
56.000	12.000			

Question 1. Nicholas claims that $\mu_x \neq 39$. Alex claims that $\mu_x = 39$. Discuss the following test at $\alpha = 0.05$:

> $H_0: \mu = 39$
> $H_1: \mu \neq 39$

Answer for #1:

a.   At $\alpha = 0.05$, $Z_{\alpha/2} = \pm 1.96$, and $Z_{st} = 0.374$, the H_0 is not rejected. Therefore, Nicholas' claim is rejected. Using the computer we can see that the lowest value of α at which the H_0 is rejected is 0.711 ($\alpha = 0.711$ = p-Value = Type I error), which implies a very low level of significance ($1 - \alpha = 0.289$). *Remember, for the H_0 hypothesis to be credibly rejected α must be equal to or less than 0.05.*

b.   Should we accept the H_0 hypothesis or Alex's claim? The answer depends on the Power of the test or on how large is the β or Type II error. *Remember, for the H_0 hypothesis to be credibly accepted β must be equal to or less than 0.05.* Let us compute a series of βs:

At $\mu_\tau = 40$,

$$\beta = P\left\{\left[\frac{(\mu_o - \mu_\tau)}{\left(\frac{s}{\sqrt{n}}\right)} - \left|Z_{\alpha/2}\right|\right] \leq Z \leq \left[\frac{(\mu_o - \mu_\tau)}{\left(\frac{s}{\sqrt{n}}\right)} + \left|Z_{\alpha/2}\right|\right]\right\}$$

$$= P\left\{\left[\frac{(39-40)}{(3.589)} - 1.96\right] \leq Z \leq \left[\frac{(39-40)}{(3.589)} + 1.96\right]\right\}$$

$$= P(-2.24 \leq Z \leq 1.68) = 0.941$$

Therefore, Power = $1-\beta = 0.059$

At $\mu_\tau = 45$,

$$\beta = P\left\{\left[\frac{(39-45)}{(3.589)} - 1.96\right] \leq Z \leq \left[\frac{(39-45)}{(3.589)} + 1.96\right]\right\}$$

$$= P(-3.63 \leq Z \leq 0.29) = 0.614$$

Therefore, Power $= 1-\beta = 0.386$

At $\mu_\tau = 32$,

$$\beta = P\left\{\left[\frac{(39-32)}{(3.589)} - 1.96\right] \leq Z \leq \left[\frac{(39-32)}{(3.589)} + 1.96\right]\right\}$$

$$= P(-0.01 \leq Z \leq 3.91) = 0.5039$$

Therefore, Power $= 1-\beta = 0.4961$

From the above it is obvious that at different values of μ_τ Alex's claim may not be accepted with high power. Thus, neither the H_0 nor the H_1 may be credibly accepted.

c. Given that $\alpha = 0.05$, and $\sigma \approx s = 20.302$, what sample size is needed so that the H_0 is credibly accepted with a β of 0.05 and a Δ of 2?

Since $\alpha = \beta = 0.05$, $Z_\alpha = Z_\beta = 1.645$.

$$\text{Therefore, } n = \frac{\left[(Z_\alpha + Z_\beta)^2 \sigma^2\right]}{\Delta^2}$$

$$= \frac{\left[(1.645 + 1.645)^2\, 20.302^2\right]}{2^2} = 1{,}115.35 \text{ or } 1{,}116.$$

Whether or not a study with a required sample size of 1,116 will be undertaken depends on sampling costs as well as the study's importance.

Question 2. Nicholas claims that $\mu_y = 15$. Alex claims that $\mu_y \neq 15$. Should they use the t or the Z distribution to test their claims?

Answer for #2:

The hypotheses are:

$H_0: \mu = 15$
$H_1: \mu \neq 15$

Because n > 30 they may use the Z or the t distribution. (The t distribution approximates the Z distribution for n > 30). The computer reports: df = 31, t_{st} = 4.74, and p-Value = 0.000045. Therefore, since the p-value is less than 0.05 the H_0 is rejected and so is Nicholas' claim.

Question 3. Nicholas claims that there is no difference in the population means of X and Y. Use the computer to test his claim with: a) a paired test; and b) an independent test. What may be concluded about the level of confidence?

Answer for #3:

The hypotheses are:

$H_0: \mu_x - \mu_y = 0$ (Nicholas' claim)
$H_1: \mu_x - \mu_y \neq 0$

a.   Paired test:

df = 31, t_{st} = 2.085, and p-value = 0.045.
Thus, reject the H_0 and the claim with 95.5% confidence.

b.   Independent test:

df = 59.9, t_{st} = 2.414, and p-value = 0.019.
Thus, reject the H_0 and the claim with 98.1% confidence.

Question 4. Alex considers the difference (D) between X and Y and he claims that D ≠ 39. Following the one-sample computer procedure, test his claim. What may be concluded about the level of confidence?

Answer for #4:

The hypotheses are:

$H_0: D = 39$
$H_1: D \neq 39$ (Alex's claim)

df = 31, t_{st} = -5.142, and p-value = 0.0000143.
Thus, reject the H_0 with 99.99% confidence and accept the claim.

Question 5. Nicholas considers the following samples for X and Y, and their statistics:

X	Y
23.000	16.000
12.000	18.000
49.000	23.000
47.000	17.000
10.000	24.000
13.000	56.000
36.000	19.000
78.000	24.000
11.000	23.000
56.000	12.000
87.000	13.000
34.000	17.000
37.000	18.000
45.000	9.000
39.000	44.000
56.000	24.000
23.000	46.000
51.000	
54.000	
23.000	

He claims that the samples do not come from the same population. Test his claim using an "independent" test.

Answer for #5:

The hypotheses are:

H_0: $\mu_x - \mu_y = 0$
H_1: $\mu_x - \mu_y \neq 0$ (Nicholas' claim)

Here we have to use the separate variances procedure because the standard deviations are substantially different from each other. Thus, df = 31.7, t_{st} = 2.699, and since p-value = 0.011 we reject the H_0 hypothesis with 98.9% confidence and accept the claim.

4.8 HEALTH AND MEDICAL QUESTIONS AND ANSWERS

Question 1. The mean serum-creatine level measured in 12 patients 24 hours after they received a newly proposed antibiotic was 1.2 mg/dL.

a. Assuming the parent population is normal, if the mean and standard deviation of serum-creatine in the general population are 1.0 and 0.4 mg/dL, respectively, then, using a 5% significance level, test if the mean serum-creatine level in this group is different from that of the general population.

b. Construct a 95% confidence interval for the true mean serum-creatine level (assuming that the standard deviation is known to be 0.4 mg/dL).

c. How is your answer from part a related to your answer from part b?

Answer for #1:

a. This is a two-tailed test with $\mu_0 = 1.0$, $\alpha = 0.05$, $1-\alpha = 0.95$, $df = n-1 = 11$, and $t_{\alpha/2} = 2.201$

$H_0: \mu = 1.0$
$H_1: \mu \neq 1.0$

$$t_{st} = \frac{\left(\overline{X} - \mu_0\right)}{\left(\dfrac{\sigma}{\sqrt{n}}\right)} = \frac{\left(1.2 - 1.0\right)}{\left(\dfrac{0.4}{\sqrt{12}}\right)} = 1.732. \text{ Thus, we fail to reject } H_0.$$

b. $P[1.2-2.201\left(\dfrac{0.4}{\sqrt{12}}\right) \leq \mu \leq 1.2+2.201\left(\dfrac{0.4}{\sqrt{12}}\right)] = 0.95$

$\Rightarrow P[0.95 \leq \mu \leq 1.45] = 0.95$

c. The 95% confidence interval contains 1.0, which is consistent with our decision to not reject H_0 at the 5% level.

Question 2. Erythromycin is a drug that has been proposed to possibly lower the risk of premature delivery. A related area of interest is its association with the incidence of side effects during pregnancy. Assume that 30% of all pregnant women complain of nausea between the 24th and 28th week of pregnancy.

Moreover, suppose that of 200 women whom are taking erythromycin regularly during this period, 110 complain of nausea. Test the hypothesis, with 95% confidence, that the incidence rate of nausea for the erythromycin group is the same as that of a typical pregnant woman.

Answer for #2:
This is a two-tailed test with $\phi_0 = 1.0$, $\alpha = 0.05$, $1-\alpha = 0.95$, and $Z_{\alpha/2} = \pm 1.96$

H_0: $\phi = 0.3$
H_1: $\phi \neq 0.3$

$$Z_{st} = \frac{(p - \phi_0)}{\sqrt{\left[\dfrac{p(1-p)}{n}\right]}} = \frac{[(110/200) - 0.3]}{\sqrt{\left[\dfrac{0.55(1-0.55)}{200}\right]}} = 7.11 > Z_{\alpha/2} = 1.96.$$

Therefore, the H_0 is rejected with 95% confidence.

Question 3. Iron-deficiency anemia is an important nutritional health problem in the United States. A dietary assessment was performed in 51 9-11-year-old male children whose families were below the poverty level. The mean daily iron intake among these children was found to be 12.50 mg with a standard deviation of 4.75 mg.

a. The standard deviation of daily iron intake in the larger population of 9-11-year-old boys was 5.56 mg. Test the hypothesis, with 95% confidence, that the variance of the low-income group is comparable to that of the general population.

b. Construct a 95% confidence interval for the true variance of daily iron intake in the low-income group. What can we infer from this confidence interval?

Answer for #3:

a. This is a two-tail test with $\alpha = 0.05$, df $= 50$, and from the χ^2 distribution: $\chi_U^2 = 67.50$ and $\chi_L^2 = 34.76$.

H_0: $\sigma^2 = 30.91$
H_1: $\sigma^2 \neq 30.91$

$$\chi^2_{st} = \frac{\left[(n-1)s^2\right]}{\sigma^2_0} = \frac{\left[(50)(22.56)\right]}{30.91} = 36.49 \text{ which is greater than } \chi^2_L \text{ and less}$$

than χ^2_U. Therefore, we do not reject H_0.

b. $P\left[\dfrac{\left[(n-1)s^2\right]}{\chi^2_L} \leq \sigma^2 \leq \dfrac{\left[(n-1)s^2\right]}{\chi^2_U}\right] = 1-\alpha$

$$P\left[\frac{\left[(51-1)22.56\right]}{34.76} \leq \sigma^2 \leq \frac{\left[(51-1)22.56\right]}{67.50}\right] = 0.95$$

$P[32.45 \leq \sigma^2 \leq 16.71] = 0.95$

Since the interval contains $\sigma^2_0 = 30.91$, the underlying variances of the low-income and general population are not significantly different.

Question 4. A possible important determinant of lung function in children is the level of cigarette smoke in the home. Suppose this question is studied by selecting two groups: Group 1 consists of 23 non-smoking children 5-9 years of age, *both* of whose parents smoke, who have a mean FEV of 2.1L and standard deviation of 0.7L; Group 2 consists of 20 non-smoking of similar age, *neither* of whose parents smoke, who have a mean FEV of 2.3L and standard deviation of 0.4L.

a. Assuming the samples are from 2 normally distributed populations, formulate the appropriate null and alternative hypotheses in this situation, then testing them at the 5% level of significance.

b. Construct a 95% confidence interval for the population mean difference in FEV between 5 to 9-year-old children whose parents smoke and comparable children whose parents do not smoke.

Answer for #4:

a.

Sample 1	Sample 2
$n_1 = 22$	$n_2 = 20$
$s_1 = 0.7$	$s_2 = 0.4$
$\overline{X}_1 = 2.1$	$\overline{X}_2 = 2.3$

$H_0: \mu_1 - \mu_2 = 0$
$H_1: \mu_1 - \mu_2 \neq 0$

$df = 22 + 20 - 2 = 40$, $t_{\alpha/2} \pm 2.021$,

and $t_{st} = (\overline{X}_1 - \overline{X}_2) \pm t_{\alpha/2} \sqrt{s_p^2 \left(\dfrac{1}{n_1} + \dfrac{1}{n_2} \right)} = -6.35$.

Since -6.35 is less than -2.021, we reject H_0. Note: See section 3.6 for formula to calculate s_p^2.

b. $(\overline{X}_1 - \overline{X}_2) \pm t_{\alpha/2} \sqrt{s_p^2 \left(\dfrac{1}{n_1} + \dfrac{1}{n_2} \right)}$

$(2.3 - 2.1) \pm 2.021 \sqrt{0.33 \left(\dfrac{1}{22} + \dfrac{1}{20} \right)}$

Question 5. The following table shows the results of a study in which researchers examined the number of car injuries and the presence of air bags in cars.

	Air Bags Present	Air Bags Not Present	Total
Car Injuries	15	24	39
No Car Injuries	30	12	42
Total	45	36	81

Test the null hypothesis, with 95% confidence that the probability of being injured and not being injured does not differ given that air bags are present in cars.

Answer for #5:
The hypotheses for this problem are:

H_0: $p_1 = p_2$
H_1: $p_1 \neq p_2$

where p_1 is the probability of a having car injuries and p_2 is the probability of not having car injuries, given that air bags are present. If the null hypothesis is true, we do not expect the probability of having and not having injuries to differ if air bags are present.

$$\chi^2_{st} = \frac{81\left[\left|(15)(12) - (24)(30)\right| - \left(\frac{81}{2}\right)\right]^2}{(45)(36)(39)(42)} = 7.62$$

Testing the null hypothesis with an $\alpha = 0.05$ and df = 1, the value from the χ^2 table is 3.84. Since $7.62 > 3.84 \Rightarrow$ we reject the null hypothesis.

Chapter 5

CORRELATION AND
LINEAR REGRESSSION

*"He who uses statistics as a drunken man
uses lampposts for support rather than illumination"*
-Andrew Lang

5.1 Correlation

The correlation that we consider here is concerned with the strength of the relationship between two normally distributed populations x and y. It is denoted by the Greek letter ρ (rho) and it varies in the interval $-1 \le \rho_{xy} \le 1$.

Three Cases:

i.   When $\rho_{xy} = 0$ variables x and y are not correlated.

ii.   When $0 \le \rho_{xy} \le 1$, variables x and y are positively correlated: an increase in x is accompanied by an increase in y and similarly a decrease in x is accompanied by a decrease in y. The closer ρ_{xy} is to 1 the stronger the positive relationship between x and y.

iii.   When $-1 \le \rho_{xy} \le 0$, variables x and y are negatively correlated: an increase in x is accompanied by a decrease in y and similarly a decrease in x is accompanied by an increase in y. The closer ρ_{xy} is to -1 the stronger the negative relationship between x and y.

Sample correlation is denoted by r_{xy} and it may be computed as follows:

$$r_{xy} = \frac{\Sigma AB}{\sqrt{(\Sigma A^2)(\Sigma B^2)}}, \text{ where } A = x - \overline{x} \text{ and } B = y - \overline{y}$$

The t-distribution, with df = n-2, may be used to test:

$H_0 : \rho = 0$ versus
$H_1 : \rho \ne 0$

The test statistic, which is good only for tests where the null hypothesis assumes a zero correlation, is:

$$t_{st(\rho)} = \frac{r_{xy}}{\sqrt{[(1 - r_{xy}^2)/(n - 2)]}},$$

where n = number of observations.

Example #1:
Data for samples x and y, drawn from normal populations, are reported in the following table:

x	3	9	4	8	2	1	2	6	4	7
y	5	6	1	2	1	1	3	5	3	8

Compute the correlation coefficient between x and y and test your claim that the population's correlation between x and y is equal to zero.

x	y	A	B	AB	A^2	B^2
3	5	-1.6	1.5	-2.4	2.56	2.25
9	6	4.4	2.5	11.0	19.36	6.25
4	1	-0.6	-2.5	1.5	0.36	6.25
8	2	3.4	-1.5	-5.1	11.56	2.25
2	1	-2.6	-2.5	6.5	6.76	6.25
1	1	-3.6	-2.5	9.0	12.96	6.25
2	3	-2.6	-0.5	1.3	6.76	0.25
6	5	1.4	1.5	2.1	1.96	2.25
4	3	-0.6	-0.5	0.3	0.36	0.25
7	8	2.4	4.5	10.8	5.76	20.25

$\overline{X} = 4.6$ and $\overline{y} = 3.5$

$\Sigma AB = 35$, $\Sigma A^2 = 68.4$, $\Sigma B^2 = 52.5$

$r_{xy} = 0.584$, and $t_{st(\rho)} = 2.033$. At the 5% level and df = 8, $t_{\alpha/2} = \pm 2.306$.

Therefore, the H_0 cannot be rejected (your claim may be valid). This correlation coefficient is not only weak but also statistically insignificant.

Example #2:
Data for samples x and y, drawn from normal populations, are reported in the following table:

x	10	16	18	17	15	13	19	10
y	12	7	3	4	6	8	2	10

Compute the correlation coefficient between x and y and test your claim that the population's correlation between x and y is equal to zero.

x	y	A	B	AB	A^2	B^2
10	12	-4.75	5.5	-26.125	22.562	30.25
16	7	1.25	0.5	0.625	1.562	0.25
18	3	3.25	-3.5	-11.365	10.562	12.25
17	4	2.25	-2.5	-5.625	5.062	6.25
15	6	0.25	-0.5	-0.125	0.062	0.25
13	8	-1.75	1.5	-2.625	3.062	2.25
19	2	4.25	-4.5	-19.125	18.062	20.25
10	10	-4.75	3.5	-16.625	22.562	12.25

$\overline{x} = 14.75$ and $\overline{y} = 6.5$

$\Sigma AB = 81$, $\Sigma A^2 = 83.5$, $\Sigma B^2 = 84.0$

$r_{xy} = -0.967$, and $t_{st(\rho)} = -9.3$. At the 5% level and df = 6, $t_{\alpha/2} = \pm 2.447$.

Therefore, the H_0 is rejected and so is your claim. This correlation coefficient is strong and statistically significant.

Example #3:
Draw scatter plots (the scatter plot is a two dimensional graph of one variable against the other) for both of the above examples. Using a statistical software package, we have:

Scatter Plot for Example #1

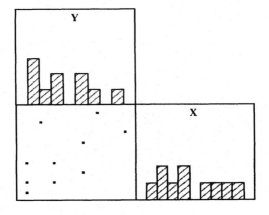

Scatter Plot for Example #2

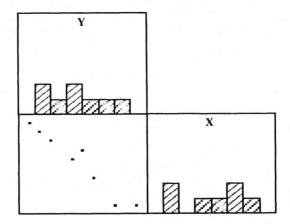

These graphs, in addition to scatter plots, display the ungrouped histograms for each variable considered, with x on the vertical axis and y on the horizontal.

5.2 Regression

Regression analysis is concerned with the dependency of a random variable (y), the dependent variable, on an independent (or explanatory) variable (x). In other words, to examine how much of the variation in y is explained by variation in x. The independent variable is assumed to have fixed values in repeated sampling. The objective of regression is to estimate and/or predict the population mean of the dependent variable given fixed values of the explanatory variable, $E(y|x)$.

5.2.1   Linear Population and Sample Regression

A linear *population* regression model is given by:

$$[1] \quad y_i = E(y|x_i) + u_I$$

where

$$[2] \quad E(y|x_i) = \beta_0 + \beta_1 x_i ,$$

and u_i = normally distributed random error, subject to certain assumptions discussed later in this chapter. (See section 5.2.4 titled Assumptions about "u"). Plugging [2] into [1] we obtain,

$$[3] \quad y_i = \beta_0 + \beta_1 x_i + u_i$$

The *sample* counterpart of [2] may be expressed as:

$$[4] \quad \ddot{y}_i = b_0 + b_1 x_i$$

where $\ddot{y}_i$ = estimator of $E(y|x_i)$, and b's = estimators of β's. The counterpart of [3] is:

$$[5] \quad y_i = b_0 + b_1 x_i + e_i$$

where e_i = sample residual term which is an estimate of u_i. Obviously, the objective in regression is to estimate

$$y_i = \beta_0 + \beta_1 x_i + u_i$$

on the basis of

$$y_i = b_0 + b_1x_i + e_i$$

5.2.2   Computation of Regression Coefficients

Another way to express equation [5] is

$$y_i = \ddot{y}_i + e_i$$

which implies

$$e_i = y_i - \ddot{y}_i$$

This last equation may be expressed as follows:

$$\Sigma(e_i)^2 = \Sigma(y_i - \ddot{y}_i)^2$$

Since $\ddot{y}_i = b_0 + b_1x_i$, we have

$$[6] \quad \Sigma(e_i)^2 = \Sigma(y_i - b_0 - b_1x_i)^2$$

Thus, it is obvious from [6] that the sum of the squared residuals is some function of b's (remember the x values are given), or $\Sigma(e_i)^2 = f(b_0, b_1)$.

Minimizing [6] with respect to b_1 and b_0 and solving simultaneously the resulting system, we can obtain estimators for both population regression coefficients. Minimization, which requires differential calculus, generates the following formulas for regression coefficients b_1 and b_0:

$$b_1 = \frac{\Sigma AB}{\Sigma A^2} \quad (\text{where } A = x - \bar{x} \text{ and } B = y - \bar{y})$$

and $b_0 = \bar{y} - b_1(\bar{x})$.

(For the calculus involved see: *Basic Econometrics*, 2nd edition, by D.N. Gujarati, McGraw Hill, 1988, Chapters 1-13).

Example #4:
Data for y = US Coffee Consumption (cups per person per day), and x = real retail price per lb. (nominal price divided by the Consumer Price Index) are given in the following table for the time period 1970-1980. [Data Source: Basic Econometrics, 2nd ed, by D.N. Gujarati, McGraw Hill, 1988, p.73]. Run a re-

gression of y against x. Using a statistical software package, graph the scatter plot and the regression line in the same diagram. Interpret the regression results.

Time	x	y
1970	0.77	2.57
1971	0.74	2.50
1972	0.72	2.35
1973	0.73	2.30
1974	0.76	2.25
1975	0.75	2.20
1976	1.08	2.11
1977	1.81	1.94
1978	1.39	1.97
1979	1.20	2.06
1980	1.17	2.02

x	y	A	B	AB	A^2
0.770	2.570	-0.241	0.364	-0.088	0.058
0.740	2.500	-0.271	0.294	-0.080	0.073
0.720	2.350	-0.291	0.144	-0.042	0.085
0.730	2.300	-0.281	0.094	-0.026	0.079
0.760	2.250	-0.251	0.044	-0.011	0.063
0.750	2.200	-0.261	-0.006	0.002	0.068
1.080	2.110	0.069	-0.096	-0.007	0.005
1.810	1.940	0.799	-0.266	-0.213	0.638
1.390	1.970	0.379	-0.236	-0.089	0.144
1.200	2.060	0.189	-0.146	-0.028	0.036
1.170	2.020	0.159	-0.186	-0.030	0.025

$\bar{x} = 1.011$ and $\bar{y} = 2.206$

$\Sigma AB = -0.611, \Sigma A^2 = 1.274$

$\therefore \quad b_1 = \dfrac{\Sigma AB}{\Sigma A^2} = \dfrac{-0.611}{1.274} = -0.47959$, and

$b_0 = \bar{y} - b_1(\bar{x}) = 2.206 - (-0.47959)(1.011) = 2.685591$, and

$y = 2.69 - 0.48x$

Scatter Plot and Regression Line

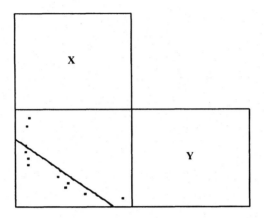

The interpretation of the regression results is as follows: If the per pound average price of coffee rises by a dollar, the average daily consumption of coffee is expected to fall by 0.48 (about half a cup). If the per pound average price of coffee rises by two dollars, the average daily consumption of coffee is expected to fall by 0.96 (about one cup). At a price of $5.604, the average daily consumption of coffee is expected to be about 0 cups. At a price of $0, the average daily consumption of coffee is expected to be about 2.69 cups!

Example #5:
Run a regression of x against y with the data in Example #1. Do the same with the data in Example #2. (Notice that the dependent variable here is x).

$$x = 2.267 + 0.667y, \text{ and } x = 21.018 - 0.964y$$

5.2.3   The Coefficient of Determination (R^2)

How well does the sample regression line fit the data? Consider Example #4. The points are scattered around the line because the fit is less than perfect. The fit would have been perfect if all the points were on the line.

To measure the goodness of fit we use the *Coefficient of Determination* or R^2 (R squared). The Coefficient of Determination measures the percentage of variation of the dependent variable explained by the regression. It varies between "0" and "1." The closer it is to "1" the better the fit and the more variation of the dependent variable is explained by the regression.

It may be computed as follows:

$$R^2 = \frac{(\Sigma AB)^2}{(\Sigma A^2)(\Sigma B^2)}$$

Important Note: The R^2 may be used to compute the correlation coefficient between y and x as follows: $r_{yx} = \pm \sqrt{R^2}$.

Example #6:
Compute the R^2 and the r_{yx} for Example #5. Interpret the results.

$$R^2 = \frac{(\Sigma AB)^2}{(\Sigma A^2)(\Sigma B^2)} = \frac{(-0.611)^2}{(1.274)(0.442)} = 0.663.$$

This result implies that only 66.3% of y's variation is explained by the regression of y against x. $r_{yx} = -\sqrt{R^2} = -0.814$. This result implies that the correlation coefficient, or the strength of the relationship, between y and x is -0.814.

5.2.4   Assumptions about the Random Error "u"

The random error, u_i, is normally distributed and subject to the following assumptions:

1.   *Y and X are linearly related.* To examine whether or not this assumption is satisfied, plot a scatter plot of Y against X. The assumption is satisfied only if the scatter plot is represented by a line.

2.   *The mean value of u_i is zero.* Or, $E(u_i|x_i) = 0$ for each i. This assumption states that the mean of u_i, given x_i, is zero. To examine the validity of this assumption, use the regression residuals (e_i) to test whether or not the mean of u_i is statistically equal to zero. If the zero hypothesis is rejected the assumption is not valid.

3.   *There is no first-order autocorrelation between the u's.* Consider the errors (u_i) and their lagged values (u_{i-1}). First-order autocorrelation is the correlation (ρ_1) of u_i with u_{i-1}. This assumption means that, given x_i, u_i and u_{i-1} are uncorrelated. To examine whether or not this assumption is satisfied one may test H_0: $\rho_1 = 0$ versus H_1: $\rho_1 \neq 0$. Rejection of the H_0 hypothesis would imply autocorrelation. Durbin and Watson have developed a statistic which may be used to make a decision. The *Durbin-Watson Statistic, d,* may be computed as follows:

$$d = \frac{\displaystyle\sum_{i=2}^{n}(e_i - e_{i-1})^2}{\displaystyle\sum_{i=1}^{n}e_i^2}$$

The d values lie within the limits $0 \le d \le 4$. If there is no autocorrelation the value of d should be about 2. In general, the critical regions of the Durbin-Watson Test are:

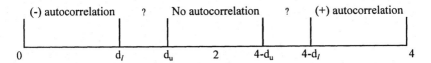

| (-) autocorrelation | ? | No autocorrelation | ? | (+) autocorrelation |

| 0 | d_l | d_u | 2 | $4-d_u$ | $4-d_l$ | 4 |

The d_l and d_u values are reported in the Durbin-Watson Table at the end of the text, for different levels of significance. Thus, only if the d value falls in the area of "no autocorrelation" the H_0: $\rho_1 = 0$ may not be rejected, which of course is our desirable result. The "?" areas imply indeterminateness.

4.   *There is no heteroscedasticity.* Var $[u_i|(x_i)] = \sigma^2$. This assumption states that the variance of u_i for each x_i is a constant number equal to σ^2. The detection of heteroscedasticity may be accomplished by using the Park Test. Park suggests running the following regression:

$$\ln e_i^2 = \alpha + \beta \ln X_i + u_i$$

If β is statistically significant, heteroscedasticity is present in the data.

Example #7:
Data for y and x are given below. Run a regression of y against x and examine the relevance of the assumptions about "u."

y	8	5	6	4	9	4	7	8	6
x	4	3	5	2	1	5	4	5	3

1. **Assumption #1:** Are y and x linearly related? Consider the scatter plot of y against x below. As it may be seen the scatter plot does not support a linear relationship. Therefore, the assumption of linearity is not satisfied.

Scatter Plot of Y against X

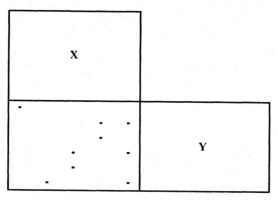

2. **Assumption #2:** Is the mean value of u_i zero? The H_0 that the mean of the regression residuals is zero may not be rejected. Therefore, the assumption is valid.

3. **Assumption #3:** Is there first-order autocorrelation between the u's? Consider the Durbin-Watson Test. The Durbin-Watson Statistic is d = 2.936. From the 5% Durbin-Watson Table we have: d_l = 0.824, d_u = 1.32. Therefore, the value of d = 2.936 falls in the area of (?), between (4-d_u) and (4-d_l). Thus, we cannot determine whether or not the data suffers from autocorrelation.

4. **Assumption #4:** Is there heteroscedasticity? (At this point we would like to ask the reader to read section 5.2.5, titled Tests of Statistical Significance). Consider the Park Test which requires the regression of (ln e^2) versus (ln x), where ln = logarithm. The regression results (with t-statistics reported in parentheses) are:

$$\ln e^2 = 1.892 - 1.518 \ln x$$
$$(1.06) \quad (-1.08)$$

As it may be seen the slope coefficient is statistically insignificant. Therefore, there is no heteroscedasticity in the data.

5.3   Tests of Statistical Significance

The statistical significance of b_0 and b_1 (corresponding to β_0 and β_1 of the population) may be examined as follows:

5.3.1   Testing of hypothesis for β_0:

Typically, we test:

H_0: $\beta_0 = 0$
H_1: $\beta_0 \neq 0$

The t-statistic is: $t_{st(\beta 0)} = \dfrac{(b_0 - \beta_0)}{\left\{ (s_e) \sqrt{\left[\left(\dfrac{1}{n} \right) + \left(\dfrac{\bar{x}^2}{ss_x} \right) \right]} \right\}}$

where $s_e = \sum \left[\dfrac{\sum (y - \ddot{y})^2}{(n-2)} \right]$ = standard error of the estimate,

and $ss_x = \sum x^2 - \left[\dfrac{(\sum x)^2}{n} \right]$

We can use the t distribution, with (n-2) degrees of freedom, to decide whether or not to reject the null hypothesis regarding β_0.

5.3.2   Testing of hypothesis for β_1:

Typically, we test:

H_0: $\beta_1 = 0$
H_1: $\beta_1 \neq 0$

The t-statistic is: $t_{st(\beta 1)} = \dfrac{(b_1 - \beta_1)}{\left[\dfrac{(s_e)}{\sqrt{ss_x}} \right]}$

As before, we can use the t distribution, with (n-2) degrees of freedom, to decide whether or not to reject the null hypothesis regarding β_1.

Important Note: A quick way to compute ss_x is to solve for it from the $t_{st(\beta1)}$ formula as follows: $ss_x = \left[\dfrac{(s_e)(t_{st(\beta1)})}{(b_1 - \beta_1)} \right]$. Statistical software provides the standard error of the estimate (s_e), the $t_{st(\beta1)}$, and the estimated coefficient b_1. The value for β_1 depends on the H_0 hypothesis. Typically, $\beta_1 = 0$.

5.3.3 Confidence intervals for regression coefficients with (n-2) degrees of freedom:

The confidence interval for β_0 is:

$$\beta_0 = b_0 \pm t_{\alpha/2} \, (s_e) \sqrt{\left[\left(\frac{1}{n} \right) + \left(\frac{\overline{x}^2}{ss_x} \right) \right]}$$

with $(1-\alpha)100$ percent confidence.

The confidence interval for β_1 is:

$$\beta_1 = b_1 \pm t_{\alpha/2} \left(\frac{s_e}{\sqrt{ss_x}} \right)$$

with $(1-\alpha)100$ percent confidence.

5.3.4 Confidence interval for mean of dependent variable when $x = x_0$ with (n-2) degrees of freedom:

The confidence interval is:

$$(\beta_0 + \beta_1 x_0) = (b_0 + b_1 x_0) \pm t_{\alpha/2} \, (s_e) \sqrt{\left[\left(\frac{1}{n} \right) + \left(\frac{\overline{x}^2}{ss_x} \right) \right]}$$

with $(1-\alpha)100$ percent confidence.

5.3.5   Interval of prediction with (n-2) degrees of freedom. This interval is not concerned with the estimation of a population parameter but instead with the prediction of a single future observation:

The interval of prediction is:

$$(\beta_0 + \beta_1 x_0) = (b_0 + b_1 x_0) \pm t_{\alpha/2}(s_e) \sqrt{\left[1 + \left(\frac{1}{n}\right) + \frac{(x_0 - \bar{x})^2}{ss_x}\right]}$$

with $(1-\alpha)100$ percent confidence.

Example #8:
Consider the following data for 12 students who have studied Japanese:

y = proficiency test scores in Japanese,
x = number of years studied.

y	60	75	80	58	73	65	89	90	70	48	52	84
x	3	4	5	2	4	3	5	5	3	2	2	4

a.   Run a regression of y against x. Compute the R^2 and interpret its meaning. Are the estimated coefficients significant at the 5% level? Why or why not? (Report the t-statistics in parentheses under the estimated coefficients). Using a statistical software package, graph the scatter plot and the regression line in the same diagram.

b.   Examine the relevance of the assumptions about "u."

c.   Use the computed regression equation to predict the proficiency score of a student who has studied Japanese for 4 years.

d.   Test at the 5% level the claim that $\beta_1 = 13$.

e.   Compute 95% confidence intervals for both estimated regression coefficients.

f.   Compute a 95% confidence interval for the mean of the dependent variable when x = 2.5.

g.   Compute an interval prediction when x = 2.5.

Answers to #8:

a. $y = 30.667 + 11.333x$
      (6.60) (8.96)

   $R^2 = 0.889$

The value of the R^2 implies that the regression explains 88.9% of y's variation. The H_0 that $\beta_0 = 0$ is rejected because at the 5% level, and n-2 = 10 degrees of freedom, the $t_{st} = 6.60 > t_{\alpha/2} = 2.228$. Thus the estimated intercept coefficient (30.667) is significant. Similarly, the H_0 that $\beta_1=0$ is rejected because the $t_{st} = 8.96 > t_{\alpha/2} = 2.228$, where $\alpha = 0.05$, with (n-2 = 10) degrees of freedom. Thus the estimated slope coefficient (11.333) is significant.

Scatter Plot and Regression Line

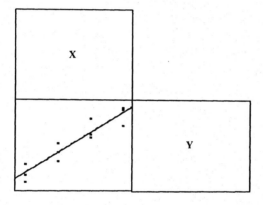

b. 1st Assumption: from the scatter plot it is obvious that the linearity assumption is satisfied.

   2nd Assumption: the H_0 that the mean of the regression residuals is zero may not be rejected. Therefore, the assumption is valid.

   3rd Assumption: The Durbin-Watson Statistic is d = 2.059. From the 5% Durbin-Watson Table we have: $d_l = 0.971$, $d_u = 1.331$. Therefore, the value of d = 2.059 falls in the area of "no autocorrelation."

   4th Assumption: The "Park" regression is:

$$\ln e^2 = 1.931 + 0.096 \ln x$$
$$(0.93) \quad (0.058)$$

Because the slope coefficient is statistically insignificant there is no heteroscedasticity in the data.

c. $y = 30.667 + 11.333(4) = 76$

d.   The H_0 that $\beta_1 = 13$ cannot be rejected because the $t_{st} = -1.32 > t_{\alpha/2} = -2.228$, where $\alpha = 0.05$, with (n-2 = 10) degrees of freedom. Thus the claim cannot be rejected.

e.   The confidence interval for β_0 is:

$$\beta_0 = b_0 \pm t_{\alpha/2}\,(s_e)\sqrt{\left[\left(\frac{1}{n}\right) + \left(\frac{\overline{x}^2}{ss_x}\right)\right]}$$

$$30.667 \pm 2.228(4.889)\sqrt{\left[\left(\frac{1}{12}\right) + \left(\frac{3.5^2}{15}\right)\right]} =$$

30.667 ± 10.35 with 95% confidence.

The confidence interval for β_1 is:

$$\beta_1 = b_1 \pm t_{\alpha/2}\left(\frac{s_e}{\sqrt{ss_x}}\right) =$$

$$11.333 \pm 2.228\left(\frac{4.889}{\sqrt{15}}\right) =$$

11.333 ± 2.81 with 95% confidence.

f.   The confidence interval for the mean of the dependent variable when $x = 2.5$ is

$$(\beta_0 + \beta_1 2.5) = (b_0 + b_1 2.5) \pm t_{\alpha/2}\,(s_e)\sqrt{\left[\left(\frac{1}{n}\right) + \left(\frac{\bar{x}^2}{ss_x}\right)\right]}$$

$= 59 \pm 10.35$ with 95% confidence.

g. The interval of prediction when $x = 2.5$ is

$$(\beta_0 + \beta_1 2.5) = (b_0 + b_1 2.5) \pm t_{\alpha/2}(s_e)\sqrt{\left[1 + \left(\frac{1}{n}\right) + \frac{(x_0 - \bar{x})^2}{ss_x}\right]}$$

$$= 59 \pm 2.228(4.899)\sqrt{\left[1 + \left(\frac{1}{12}\right) + \frac{(2.5 - 3.5)^2}{15}\right]}$$

$= 59 \pm 11.71$ with 95% confidence.

5.4   GENERAL QUESTIONS AND ANSWERS

Question 1. Consider the following data for product z: where p = price per unit sold, and q = number of units sold.

p	2	5	6	6	7	10	15	12	9	5	6	8	11	12
q	20	19	18	19	15	13	5	9	10	13	12	9	7	5

a. Run a regression of q against p. Compute the R^2 and interpret its meaning. Are the estimated coefficients significant at the 5% level? Why or why not? (Report the t-statistics in parentheses under the estimated coefficients). Graph the scatter plot and the regression line in the same diagram.
b. Examine the relevance of the assumptions about "u."
c. Compute the correlation coefficient between p and q and test its statistical significance at the 5% level.
d. Use the computed regression equation to predict the number of units sold at a price of 3.
e. Test at the 5% level the claim that $\beta_1 = -1.5$.
f. Compute 95% confidence intervals for both estimated regression coefficients.
g. Compute a 95% confidence interval for the mean of the dependent variable when p = 3.
h. Compute an interval prediction when p = 3.

Answers for #1:

a. q = 27.88 - 2.08p
      (16.15)   (-9.35)

   $R^2 = 0.879$

The value of the R^2 implies that the regression explains 87.9% of p's variation. The H_0 that $\beta_0 = 0$ is rejected because at the 5% level, and n-2 = 12 degrees of freedom, the $t_{st} = 16.15 > t_{\alpha/2} = 2.179$. Thus the estimated intercept coefficient (27.88) is significant.

Similarly, the H_0 that $\beta_1 = 0$ is rejected because at the 5% level and n-2 = 12 degrees of freedom the $t_{st} = -9.35 < t_{\alpha/2} = -2.179$. Thus the estimated slope coefficient (-2.08) is significant.

Scatter Plot and Regression Line

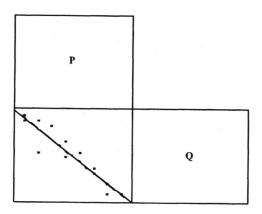

b. 1st Assumption: from the scatter plot it is obvious that the linearity assumption is satisfied.

   2nd Assumption: the H_0 that the mean of the regression residuals is zero may not be rejected. Therefore, the assumption is valid.

   3rd Assumption: The Durbin-Watson Statistic is $d = 1.802$. From the 5% Durbin-Watson Table we have: $d_l = 1.05$, $d_u = 1.35$. Therefore, the value of $d=1.802$ falls in the area of "no autocorrelation."

   4th Assumption: The "Park" regression is:

   $$\ln e^2 = 4.226 - 2.426 \ln p$$
   $$(1.03) \quad (-1.16)$$

Because the slope coefficient is statistically insignificant there is no heteroscedasticity in the data.

c. The Correlation Coefficient is $r_{pq} = -0.94$.
   At the 5% level and with $n-2 = 12$ degrees of freedom, the t statistic is

   $$t_{st(\rho)} = \frac{r_{pq}}{\sqrt{\left[\frac{\left(1 - r_{pq}^2\right)}{(n-2)}\right]}} = -9.54 < t_{\alpha/2} = -2.179.$$

Thus, the estimated correlation coefficient is significant.

d. $q = 27.88 - 2.08(3) = 21.64$

e.   The H_0 that $\beta_1 = -1.5$ is rejected because at the 5% level and 12 degrees of freedom the $t_{st} = -2.61 < t_{\alpha/2} = -2.179$. Thus, the claim is rejected.

f.   The confidence interval for β_0 is:

$$\beta_0 = b_0 \pm t_{\alpha/2}\,(s_e) \sqrt{\left[\left(\frac{1}{n}\right) + \left(\frac{\bar{x}^2}{ss_x}\right)\right]} =$$

$$27.88 \pm 2.179(1.88) \sqrt{\left[\left(\frac{1}{14}\right) + \left(\frac{7.43^2}{71.43}\right)\right]} =$$

27.88 ± 3.76 with 95% confidence.

The confidence interval for β_1 is:

$$\beta_1 = b_1 \pm t_{\alpha/2} \left(\frac{s_e}{\sqrt{ss_p}}\right) =$$

$$-2.08 \pm 2.179 \left(\frac{1.88}{\sqrt{71.43}}\right) =$$

-2.08 ± 0.484 with 95% confidence.

g.   The confidence interval for the mean of the dependent variable when $p = 3$ is

$$(\beta_0 + \beta_1 3) = (b_0 + b_1 3) \pm t_{\alpha/2}(s_e) \sqrt{\left[1 + \left(\frac{1}{n}\right) + \frac{(3 - \bar{p})^2}{ss_p}\right]} =$$

$$21.64 \pm 2.179\,(1.88)\sqrt{\left[1+\left(\frac{1}{14}\right)+\frac{(3-7.43)^2}{71.43}\right]} =$$

21.64 ± 2.41 with 95% confidence.

h. The interval of prediction when $p = 3$ is

$$(\beta_0+\beta_1 3) = (b_0+b_1 3) \pm t_{\alpha/2}(s_e)\sqrt{\left[1+\left(\frac{1}{n}\right)+\frac{(p-\bar{p})^2}{ss_p}\right]} =$$

$$21.64 \pm 2.179\,(1.88)\sqrt{\left[1+\left(\frac{1}{14}\right)+\frac{(3-7.43)^2}{71.43}\right]} =$$

21.64 ± 4.75 with 95% confidence.

Question 2. Consider the following data for product k: where c = cost per unit of production, and q = number of units produced.

c	2	5	6	6	7	10	15	12	9	5	6	8	11	12
q	12	15	16	16	18	20	24	19	17	14	15	17	21	23

a. Compute the regression of c versus q, the t-Statistics for the estimated coefficients, the R^2, the Durbin-Watson Statistic, the "Park" regression, and graph the scatter plot.

b. Discuss the findings. Should we use the regression for inference?

c. Discuss possible modifications.

Answers for #2:

a. The regression results are:

$$c = -2.836 + 0.582q$$
$$(-1.53) \quad (5.64)$$

$R^2 = 0.726$

Durbin-Watson Statistic: $d = 1.166$

"Park" regression: $\ln e^2 = -7.11 + 2.308 \ln q$
$\qquad\qquad\qquad\qquad (-1.15) \quad (1.07)$

S catter Plot and Regression Line

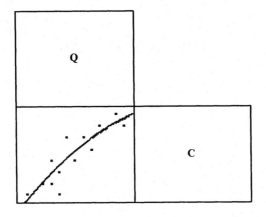

b. The slope coefficient is significant, there is no heteroscedasticity, and the R^2 is bearable. The 1st, 2nd, and 3rd assumptions though are violated: the relationship appears nonlinear, the H_0 that the mean of the residuals is zero is rejected at the 5% level, and d falls in the "question mark" area. On the basis of these results we should not use this regression for inference.

c. Proper modifications would be to collect more data for c and q, re-specify the model to accommodate for non-linearity (e.g. assume that the population regression is $c = \alpha q^\beta$), measure differently c and q, etc. (For more on such modifications see the Gujarati book, mentioned above).

Question 3. Consider the data for y and x reported in the following table.

y	7	8	4	6	1	7	9	3	5	6
x	16	17	12	13	10	17	20	11	15	16

a. Run a regression of y against x and discuss the findings.
b. Compute the correlation coefficient and discuss its statistical significance.

c. Compute a 99% prediction interval for y, given x = 14.

Answers for #3:

a. The regression results are:

$$y = -5.045 + 0.724 \, x$$
$$\quad\;\; (0.007) \;\; (0.000056)$$

$$R^2 = 0.882$$

Durbin-Watson Statistic: d = 1.918

"Park" Regression: $\ln e^2 = -1.929 + 0.203 \ln x$
$$\quad\quad\quad\quad\quad\quad (0.797) \;\; (0.947)$$

Scatter Plot and Regression Line

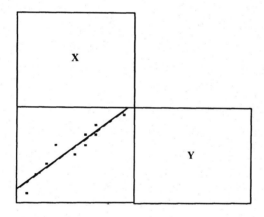

As it may be seen none of the assumptions is violated: the relationship appears linear, the H_0 that the mean of the residuals is zero cannot be rejected at the 5% level (p-value = 0.993), d falls in the area of no autocorrelation, and there is no heteroscedasticity. The R^2 implies that 88.2% of y's variation is explained by the regression, a value that could be improved with more data, better specification of the relationship between y and x, and other such modifications. The estimated coefficients are highly significant: the H_0 hypotheses for β_0 and β_1 are both rejected, since the p-values (reported in parentheses) are both less than 0.05. On the basis of these results we may safely use this regression for inference.

b. $r_{YX} = 0.939$, p-value = 0.00005624.

Thus, the relationship between Y and X is positive, very strong, and highly significant since the p-value is less than 0.05.

c. $s_e = 0.88$, $t_{st(\beta 1)} = 7.723$, $b_1 = 0.724$, and $\overline{X} = 14.7$.

Thus, since $\beta_1 = 0$, $ss_x = [(s_e)(t_{st(\beta 1)})/(b_1-\beta_1)]^2 = 88.12$.

From the t table, with n-2 degrees of freedom, $t_{\alpha/2} = 3.355$.

Therefore, $y = 5.09 \pm 3.104$, with 99% confidence.

5.5 HEALTH AND MEDICAL QUESTIONS AND ANSWERS

Question 1. Torok-Storb et al. (1985) examined 9 patients with Aplastic Anemia, hypothesizing that variation in the number of lymphocytes can be explained by the percentage of reticulytes. The data employed in the study is listed in the following table.

Patient	% Reticulytes	Lymphocytes (per mm^2)
1	3.6	1,700
2	2.0	3,078
3	0.3	1,820
4	0.3	2,706
5	0.2	2,086
6	3.0	2,299
7	0.0	676
8	1.0	2,088
9	2.2	2,013

a. Compute the regression line relating the percentage of reticulytes (x) to the number of lymphocytes (y).
b. What is the R^2 for the regression line? Interpret this value.
c. Test the statistical significance of the b_1 coefficient using a t-test at the 5% level.

Answer for #1:

a. $y = 1,894.8 + 112.1x$

b. 0.05. R^2 indicates the percentage of lymphocyte count that is explained by the percentage variation of reticulytes.

c.

$H_0 : b_1 = 0$ versus
$H_1 : b_1 \neq 0$

Since $t_b = 0.61 < t_{st} = 2.306 \Rightarrow$ Do not reject H_0

Question 2. The National Center for Heath Statistics (1979) reported the infant-mortality rates per 1,000 live births in the United States for the period 1960-1979 as

Year (x)	Infant-Mortality Rate (y)
1960	26.0
1965	24.7
1970	20.0
1971	19.1
1972	18.5
1973	17.7
1974	16.7
1975	16.1
1976	15.2
1977	14.1
1978	13.8
1979	13.0

a. Compute the regression line relating the year (x) to the infant-mortality rate (y).
b. What is the R^2 for the regression line? Interpret this value.
c. Test the statistical significance of the b_1 coefficient using a t-test at the 5% level.

Answers for #2:

a. $y = 1471.97 - 0.737x$

b. 0.98. R^2 indicates that 98% of the variation in infant-mortality is explained by time.

c.

$H_0: b_1 = 0$ versus
$H_1: b_1 \neq 0$

Since $t_b = -23.53 < t_{st} = -2.201 \Rightarrow$ We reject H_0

Question 3. The Second Task Force on Blood Pressure Control in Children (1987) reported the observed 90th percentile of systolic blood pressure (SBP) in single years of age from age 1 to 18 based on prior studies. The data for boys is:

Age (x)	SBP (y)
1	105
2	106
3	107
4	108
5	109
6	111
7	112
8	114
9	115
10	117
11	119
12	121
13	124
14	126
15	129
16	131
17	134
18	136

a. Using a statistical software package, fit a regression line relating age (x) to SBP (y).
b. What is the R^2 for the regression line? Interpret this value.
c. Test the statistical significance of the b_1 coefficient using a t-test at the 5% level.

Answers for #3:

a. $y = 100.39 + 1.85x$

b. 0.98. R^2 indicates that 98% of the variation in SBP is explained by age.

c. H_0: $b_1 = 0$ versus
    H_1: $b_1 \neq 0$

Since $t_b = 25.30 > t_{st} = 2.110 \Rightarrow$ We reject H_0

Chapter 6

NONPARAMETRIC TESTS

"Make everything as simple as possible, but no simpler"
-Albert Einstein

Our methods of estimation and hypothesis testing, so far, are based on the assumption that parent populations are approximately normally distributed or that the *parametric* forms of populations are known. If this assumption cannot be made, and if the central-limit theorem cannot be invoked, then *nonparametric* or *distribution-free testing* must be used.

Distribution-free tests can be used when: (a) the parent population has an unknown shape, (b) the central-limit theorem cannot be applied, (c) after testing for normality, we are in doubt about the shape of the parent population, and (d) we have ordinal data - data that can be ranked on the basis of importance.

Many nonparametric statistics are invariant under rank order transformations of the data points. In other words, we may change actual data values as long as we preserve relative ranks, and the results of our hypothesis tests will not change. Data that can be replaced by rank order values without losing information are often called rank or *ordinal data*. For example, if we believe the list (-2, 45, 119.6, 2456) contains only ordinal information, then we can replace it with the list (1, 2, 3, 4) without loss of information.

Some nonparametric methods are invariant under permutation transformations. That is, you can interchange data values and get the same results, provided you keep all data points with one value before transformation single valued after transformation. Data that you can treat like this are often called *nominal* or *categorical data*. For example, if we believe the list (3, 3, 6, 6, 9, 9, 9) contains only nominal information, then we can replace it with the list (black, black, white, white, yellow, yellow, yellow) without loss of information.

Distribution-free tests should not be used when parametric testing is possible. Parametric testing is: (a) more powerful (lower type II error) than nonparametric testing under the same conditions, and (b) more efficient since it does not ignore available sample information: for instance, a nonparametric test focuses only on the direction rather than also on sizes of differences between data points.

How sensitive are tests to deviations from basic assumptions? Some tests are very robust: their degree of sensitivity to errors in assumptions is low. Other tests are not so robust. If we can afford to lose some power by using a nonparametric test, we can gain *robustness*.

In whatever follows we utilize a statistical software package to introduce the *Lilliefors* test for normality, and then some of the most popular nonparametric tests. Namely, these tests are: the *Wilcoxon Rank-Sum* or *Mann-Whitney*, the *Sign*, the *Wilcoxon Signed-Rank*, the *Kruskal-Wallis*, and the *Spearman's Rank-Correlation*.

6.1 The Lilliefors Test for Normality

We can test normality without assuming a particular mean or standard deviation for the distribution by performing the Lilliefors (1967) test. The procedure automatically standardizes the variables and tests whether the standardized versions are normally distributed. The test is not affected by the variable's original mean and standard deviation. The zero and alternative hypotheses are:

H_0: The parent population may be normally distributed
H_1: The parent population is not normally distributed

If the p-value is less than 0.05 reject the zero hypothesis.

Example #1:
Data for samples G and L are given in the following table. Examine if the samples can "pass" normality.

G	11	11	8	14	11.5	18	17	17	13	19	13	14	15
L	20	20	18	20	20	20	19	19	18	20	20	20	20

The computer results are:

KOLMOGOROV-SMIRNOV ONE SAMPLE TEST
USING STANDARD NORMAL DISTRIBUTION

VARIABLE	N-OF-CASES	LILLIEFORS PROBABILITY (2-TAIL)
G	13.000	0.842
L	13.000	0.000

The zero hypothesis for G may not be rejected and therefore its parent population may be normal. The zero hypothesis for L is rejected and therefore its parent population is not normal.

6.2 The Wilcoxon Rank-Sum or Mann-Whitney Test

This test determines whether two statistical populations of *continuous* values are identical to or different from one another, and it is based on *independent* samples.

Procedure:

A. Consider the following hypotheses:

H_0: The distributions of the parent populations are identical.
H_1: The distributions of the parent populations differ.

B. Subject the samples to the Lilliefors test. If any one of them does not pass normality, follow the steps below. If both samples pass normality, utilize parametric methods.

C. To derive the test statistic that is needed to run this test we have to:

i. Combine the data contained in two independent samples of sizes n_a and n_b. Sample sizes need not be of equal size but each must be equal or greater than 10.

ii. Rank the pooled data from smallest (call it 1) to largest value. For two or more identical data values use their average rank. For example, let us assume that we have to rank the numbers 3, 8, 9, 5, 4, and 8. Arranging these numbers from lowest to highest we have 3, 4, 5, 8, 8, 9. Because numbers "8" are in the 4th and 5th spots their average ranking is $(4+5)/2 = 4.5$. Thus, the rankings for all numbers are: 1, 2, 3, 4.5, 4.5, 6;

iii. Sum the ranks in each sample and designate either one of these rank sums (R_a or R_b) as the test statistic.

D. For sample a, compute

$$\mu_{R_a} = \frac{n_a (n_a + n_b + 1)}{2}$$

$$\sigma_{R_a} = \sqrt{\frac{\left[\dfrac{(n_a n_b)}{(n_a + n_b + 1)} \right]}{12}}$$

$$Z = \frac{\left(R_a - \mu_{R_a}\right)}{\sigma_{R_a}}$$

If Z falls between $\pm Z_{\alpha/2}$ the H_0 hypothesis may not be rejected. Otherwise, the H_0 hypothesis is rejected. (For sample b, interchange subscripts a and b in the above formulas.)

Example #1:
Consider the following data for samples a and b:

a	32	45	34	56	45	57	67	26	20	16	15		
b	12	20	78	93	59	95	110	245	300	456	1,123	4,456	5,679

Test our claim that the samples do not come from identical populations.

Steps:
A. We are interested in testing the following hypotheses:

H_0: a's and b's populations are identical.
H_1: a's and b's populations differ.

Our claim is identical to the alternative hypothesis.

B. The computer results for the Lilliefors test are:

KOLMOGOROV-SMIRNOV ONE SAMPLE TEST
USING STANDARD NORMAL DISTRIBUTION

VARIABLE	N-OF-CASES	LILLIEFORS PROBABILITY (2-TAIL)
a	11.000	1.000
b	13.000	0.000

According to these results the zero hypothesis that the population may be normal is rejected only for sample b. Therefore, the use of nonparametric analysis is justified.

C. The following table lists the pooled data sorted from lowest to highest (column 1), sample of origin (column 2), rank (column 3), a's ranks (column 4), and b's ranks (column 5).

Data (1)	Sample (2)	Rank (3)	a's Ranks (4)	b's Ranks (5)
12	"b"	1.0		1
15	"a"	2.0	2	
16	"a"	3.0	3	
20	"a"	4.5	4.5	
20	"b"	4.5		4.5
26	"a"	6.0	6	
32	"a"	7.0	7	
34	"a"	8.0	8	
45	"a"	9.5	9.5	
45	"a"	9.5	9.5	
56	"a"	11.0	11	
57	"a"	12.0	12	
59	"b"	13.0		13
67	"a"	14.0	14	
78	"b"	15.0		15
93	"b"	16.0		16
95	"b"	17.0		17
110	"b"	18.0		18
245	"b"	19.0		19
300	"b"	20.0		20
456	"b"	21.0		21
1123	"b"	22.0		22
4456	"b"	23.0		23
5679	"b"	24.0		24
			$R_a = 86.7$	$R_b = 213.5$

D. Thus, using sample a:

$$\mu_{R_a} = \frac{n_a\left(n_a + n_b + 1\right)}{2} = 137.5$$

$$\sigma_{R_a} = \sqrt{\frac{\left[\frac{(n_a n_b)}{(n_a + n_b + 1)}\right]}{12}} = 17.26$$

$$Z = \frac{\left(R_a - \mu_{R_a}\right)}{\sigma_{R_a}} = 2.94$$

Therefore at the 5% level $Z < -1.96$ and the H_0 hypothesis is rejected. (Similarly with sample b)

The computer results are:

KRUSKAL-WALLIS ONE-WAY ANALYSIS
OF VARIANCE FOR 24 CASES

DEPENDENT VARIABLE IS   AB
GROUPING VARIABLE IS   AB$

GROUP	COUNT	RANK SUM
b	13	213.500
a	11	86.500

MANN-WHITNEY U TEST STATISTIC =   122.500
PROBABILITY IS   0.003
CHI-SQUARE APPROXIMATION =   8.738 WITH 1 DF

As we see the statistical software gives us the sums, the p-value, plus the Mann-Whitney U test statistic, and the X^2 approximation. The p-value = 0.003 indicates that the H_0 is rejected not only at the 5% level but also at the 1%.

6.3 The Sign Test

This test determines whether two statistical populations of *continuous* values are identical to or different from one another, and it is based on a *matched-pairs* sample. This test takes into consideration only the sign of the difference between the data cases in a matched pair. The number of positive signs among all the pairs is used as the test statistic. Matched pairs with a zero difference are omitted.

Thus, n = (number of positive signs) + (number of negative signs).

Procedure:

A. Consider the following hypotheses:

H_0: The distributions of the parent populations are identical.
H_1: The distributions of the parent populations differ.

B. i.   Quantitative measures of matched-pairs are available: subject the samples to the Lilliefors test. If any one of them does not pass normality follow the steps below. If both samples pass normality utilize parametric methods.

ii. Quantitative measures of matched-pairs are not available: follow the steps below.

C. Record the sign of difference between the matched pairs of the data points. The number of positive signs is the test statistic (S).

D. Compute

$\mu_s = 0.5n$, where n = (# of "+" signs) + (# of "-" signs),

$\sigma_s = \sqrt{0.25n}$, and for $n \geq 10$

$$Z = \frac{(S - \mu_s)}{\sigma_s}.$$

If Z falls between $\pm Z_{\alpha/2}$ the H_0 hypothesis may not be rejected. Otherwise, the H_0 hypothesis is rejected.

Example #1:
Consider the following data for samples a and b:

a	19	18	19	9	19	18	19	19	7	18	18	19	19	18
b	13	11	16	7	14	13	10	11	8	15	18	18	22	19

Test our claim that the samples come from identical populations.

Steps:
A. We are interested in testing the following hypotheses:

H_0: a's and b's populations are identical.
H_1: a's and b's populations differ.

Our claim is identical to the null hypothesis.

B. The Lilliefors test indicates that we cannot use parametric methods.

C. Sign of Difference (a-b):

a	19	18	19	9	19	18	19	19	7	18	18	19	19	18
b	13	11	16	7	14	13	10	11	8	15	18	18	22	19
a-b	+	+	+	+	+	+	+	+	-	+	0	+	-	-

The number of positive signs S = 10.

D. Thus,

$$n = 13 \text{ and } \mu_s = 0.5n = 6.5,$$

$$\sigma_s = \sqrt{0.25n} = 1.803, \text{ and}$$

$$Z = \frac{(S - \mu_s)}{\sigma_s} = 1.94.$$

Therefore, at the 5% level $Z < 1.96$ and the H_0 hypothesis may not be rejected which implies that our claim may be valid.

The computer results are:

SIGN TEST RESULTS
COUNTS OF DIFFERENCES
(ROW VARIABLE GREATER THAN COLUMN)

	a	b
a	0	10
b	3	0

TWO-SIDED PROBABILITIES FOR EACH PAIR OF VARIABLES

	a	b
a	1.000	
b	0.092	1.000

As we see, the statistical software gives us the counts of differences (10 and 3) and the p-value (0.092), which indicates that the zero hypothesis is rejected at the 9.2% level.

Example #2:
A toothpaste manufacturer is about to launch a new marketing campaign for a new brand. The company's production department has come up with two potential brands, but due to financial restrains, the company can market only one of them, the one for which the market would show preference. To determine whether brand x or brand y would be marketed, the company selected randomly 100 people to test the brands. After a week, 51 people reported that they liked x more than y, 3 found x and y equally good, and the remaining 46 found y better than x. Should the company market x or y?

We are interested in testing the following hypotheses:

H_0: Consumers are indifferent between x and y.
H_1: Consumers are not indifferent between x and y.

The Lilliefors test may not be applied because we do not have quantitative information about x and y.

The number of positive signs (x better than y) S = 51.

Thus,

$$n = 97 \text{ and } \mu_s = 0.5n = 48.5,$$

$$\sigma_s = \sqrt{0.25n} = 4.92, \text{ and}$$

$$Z = \frac{(S - \mu_s)}{\sigma_s} = 0.51$$

Therefore, at the 5% level Z < 1.96 and the H_0 hypothesis may not be rejected which implies that consumers are indifferent between x and y.

We may not use statistical software for this problem because quantitative data for x and y does not exist.

Example #3:
A realtor claims that the median house price in her town is $65,000. To test her claim, a sample of 120 houses for sale is randomly selected from a real estate listing. 64 houses are for sale at a price higher than $65,000. Is the realtor's claim valid?

In this example, variable x consists of 120 numbers all equal to $65,000. Variable y consists of 64 numbers greater than $65,000 and 56 lower than $65,000. We are interested in testing the following hypotheses:

H₀: The population's median is $65,000.
H₁: The population's median is not $65,000.

H_0: The population's median is $65,000.
H_1: The population's median is not $65,000.

The number of positive signs S = 64. Thus,

$$n = 120 \text{ and } \mu_s = 0.5n = 60,$$

$$\sigma_s = \sqrt{0.25n} = 5.477, \text{ and}$$

$$Z = \frac{(S - \mu_s)}{\sigma_S} = 0.73$$

Therefore, at the 5% level Z < 1.96 and the H₀ hypothesis, that the median is $65,000, may not be rejected.

6.4 The Wilcoxon Signed-Rank Test

This test determines whether two statistical populations of *continuous* values are identical to or different from one another, and it is based on a *matched-pairs* sample. This test takes into consideration not only the sign of the difference between the data cases in a matched pair but also the magnitude of their difference. This test is a more powerful alternative to the sign test.

Procedure:

A. Consider the following hypotheses:

H_0: The distributions of the parent populations are identical.
H_1: The distributions of the parent populations differ.

B. Subject the samples to the Lilliefors test. If any one of them does not pass normality follow the steps below. If both samples pass normality utilize parametric methods.

C. To derive the test statistic that is needed to run this test we have to:

i. Calculate the difference between each matched pair.
Let n = number of pairs with nonzero differences.

ii. Rank the absolute values of the n numbers from smallest (call it 1) to largest.

iii. Attach to each rank the sign of the original difference that corresponds to it and compute the sum (T) of these ranks. T is the test statistic.

D.

$$\mu_T = 0$$

$$\sigma_T = \sqrt{\frac{[n(n-1)(2n-1)]}{6}} \text{ , and if } n \geq 10$$

$$Z = \frac{(T - \mu_T)}{\sigma_T}$$

If Z falls between $\pm Z_{\alpha/2}$ the H_0 hypothesis may not be rejected. Otherwise, the H_0 hypothesis is rejected.

Example #1:
Consider Example #3 in the previous section.

Steps:

A. We are concerned with the following hypotheses:

H_0: The distributions of the parent populations are identical.
H_1: The distributions of the parent populations differ.

B. The samples, as explained in Example #3, do not pass the Lilliefors test for normality.

C.

a	b	a-b	|a-b|	Rank of |a-b|	Signed Rank
19.0	13.0	6.0	6.0	10.0	10.0
18.0	11.0	7.0	7.0	11.0	11.0
19.0	16.0	3.0	3.0	6.0	6.0
9.0	7.0	2.0	2.0	4.0	4.0
19.0	14.0	5.0	5.0	8.5	8.5
18.0	13.0	5.0	5.0	8.5	8.5
19.0	10.0	9.0	9.0	13.0	13.0
19.0	11.0	8.0	8.0	12.0	12.0
7.0	8.0	-1.0	1.0	2.0	-2.0
18.0	15.0	3.0	3.0	6.0	6.0
18.0	18.0	0.0	0.0	.	.
19.0	18.0	1.0	1.0	2.0	2.0
19.0	22.0	-3.0	3.0	6.0	-6.0
18.0	19.0	-1.0	1.0	2.0	-2.0

$$T = 71$$

D.

$$\mu_T = 0$$

$$\sigma_T = \sqrt{\frac{[n(n-1)(2n-1)]}{6}} = 70.1, \text{ and since } n > 10,$$

$$Z = \frac{(T - \mu_T)}{\sigma_T} = 1.013$$

At the 5% level $Z < 1.96$. Therefore, the H_0 hypothesis may not be rejected.

The computer results are:

WILCOXON SIGNED RANKS TEST RESULTS

COUNTS OF DIFFERENCES
(ROW VARIABLE GREATER THAN COLUMN)

	a	b
a	0	10
b	3	0

Z = (SUM OF SIGNED RANKS)

SQUARE ROOT(SUM OF SQUARED RANKS)

	<u>a</u>	<u>b</u>
a	0.000	
b	-2.488	0.000

TWO-SIDED PROBABILITIES USING NORMAL APPROXIMATION

	<u>a</u>	<u>b</u>
a	1.000	
b	0.013	1.000

As we see the statistical software gives us the counts of differences (10 and 3), the z statistic (computed differently), and the p-value (0.013), which indicates that the zero hypothesis is rejected at the 2% level.

Example #2:

A factory produces electric bulbs with two machines, a and b. To examine whether or not the machines produce identical bulbs, samples from a and b are collected and the burning days of each bulb are recorded. The samples, in terms of burning days, are:

a	245	345	167	200	256	268	256	279	258	245	198	178	145
b	363	363	362	345	333	300	363	365	366	364	200	198	168

The production manager claims that the machines produce different qualities of bulbs. Test the manager's claim. (Use the computer.)

The relevant hypotheses are:

H_0: The machines are identical (or, the samples come from identical populations)

H_1: The machines are not identical (or, the samples do not come from identical populations)

Using the statistical software, we see that for b the Lilliefors test indicates non-normality. The Wilcoxon Signed-Rank test computer results are:

WILCOXON SIGNED RANKS TEST RESULTS

COUNTS OF DIFFERENCES
(ROW VARIABLE GREATER THAN COLUMN)

	ba	bb
ba	0	0
bb	13	0

Z = (SUM OF SIGNED RANKS)
SQUARE ROOT(SUM OF SQUARED RANKS)

	ba	bb
ba	0.000	
bb	3.180	0.000

TWO-SIDED PROBABILITIES USING NORMAL APPROXIMATION

	ba	bb
ba	1.000	
bb	0.001	1.000

The P-value (0.001) indicates that the H_0 should be rejected. Therefore, the manager's claim, that the machines are not identical, is valid.

6.5 The Kruskal-Wallis Test

This test determines whether *more than two* statistical populations of *continuous* values are identical to or different from one another, and it is based on *independent* samples.

Procedure:

A. The Hypotheses tested are:

H_0: The samples come from identical populations
H_1: The samples do not come from identical populations

B. Subject the samples to the Lilliefors test. If any one of them does not pass normality follow the steps below. If both samples pass normality utilize parametric methods.

C. This test is an extension of the Wilcoxon Rank-Sum test, and as in that test, the data points contained in the samples are pooled and ranked. A rank sum is computed for each original sample and the Kruskal-Wallis Test Statistic as follows:

$$K = \left[\frac{12}{n(n-1)}\right]\left[\Sigma\left(\frac{W_i^2}{n_i}\right)\right] - 3(n+1)$$

where,

n = number of observations in all samples,
W_i = rank sum of an individual sample,
n_i = number of observations in that sample.

D. Use the χ^2 distribution with Degrees of Freedom (DF), DF = n_s-1, where n_s = number of samples. If $K \leq \chi^2_{\alpha,DF}$ the H_0 may not be rejected.

Example #1:
The manufacturer of a certain machine claims that all of his machines produce products of the same quality. To test his claim the buyer selected randomly three machines and let each machine produce 100 items. The number of non-defective items was used as a measure of quality, and the experiment was repeated 5 times. The number of non-defective items per run is reported in the following table:

Run	Machine A	Machine B	Machine C
1	78	89	95
2	88	92	79
3	94	96	82
4	85	93	86
5	91	90	98

The relevant hypotheses are:

H_0: The machines produce products of identical quality (or, the samples come from identical populations)
H_1: The machines do not produce products of identical quality (or, the samples do not come from identical populations)

The ranks of the data (where A_r, B_r, C_r are the ranks of A, B, and C respectively) are:

A	Ar	B	Br	C	Cr
78	1	89	7	95	13
88	6	92	10	79	2
94	12	96	14	82	3
85	4	93	11	86	5
91	9	90	8	98	15
Rank Sums:	32		50		38

At $\alpha = 0.05$, and DF = 2, K = $1.68 < \chi^2_{.05,2} = 5.991$ the H_0 may not be rejected. Therefore, the manufacturer's claim may not be rejected.

We may also use a statistical software package as follows: start a new file, type in the same column the pooled data. Then, in a column next to it, create a grouping variable that consists of letters that correspond to the sample points. For instance, in Example #7 the pooled data (PD) and the grouping variable (PD$) must be entered as follows:

PD	78	88	94	85	91	89	92	96	93	90	95	79	82
PD$	A	A	A	A	A	B	B	B	B	B	C	C	C

The computer results are:

KRUSKAL-WALLIS ONE-WAY ANALYSIS
OF VARIANCE FOR 15 CASES

DEPENDENT VARIABLE IS   PD
GROUPING VARIABLE IS   PD$

GROUP	COUNT	RANK SUM
A	5	32.000
B	5	50.000
C	5	38.000

KRUSKAL-WALLIS TEST STATISTIC = 1.680
PROBABILITY IS 0.432 ASSUMING CHI-SQUARE DISTRIBUTION WITH 2 DF

6.6 The Spearman's Rank-Correlation Test

This test measures the correlation (strength of relationship) between two rank-ordered variables. The test statistic, called Spearman Rank-Correlation Coefficient (ρ), is:

$$\rho = 1 - \left[\frac{6\left(\sum d_i^2\right)}{n(n^2 - 1)} \right]$$

$d_i = x_i - y_i$, where x_i (y_i) = ranking of data point i of variable $X(Y)$.

$-1 \le \rho \le 1$

Procedure:

A. Consider the following hypotheses:

H_0: The X and Y rankings are not correlated
H_1: The X and Y rankings are correlated

B. Compute

$$\mu_\rho = 0,$$

$$\sigma_\rho = \sqrt{\frac{1}{(n-1)}} \text{, and for } n \ge 10$$

$$Z = \frac{\left(\rho - \mu_\rho\right)}{\sigma_\rho}$$

If Z falls between $\pm Z_{\alpha/2}$ the H_0 hypothesis may not be rejected. Otherwise, the H_0 hypothesis is rejected.

Example #1:
Baseball teams a, b, c, d, e, f, g, h, i, j, k, l, m, and n, have been ranked by two different groups of judges as follows:

Team	a	b	c	d	e	f	g	h	i	j	k	l	m	n
Grp X	1	2	3	4	5	6	7	8	9	10	11	12	13	14
Grp Y	2	1	3	5	4	6	8	9	11	10	12	7	13	14

Does a significant rank correlation exist between X and Y?

The relevant hypotheses are:

H_0: The X and Y rankings are not correlated
H_1: The X and Y rankings are correlated

The difference (d) between X and Y, d^2, and the sum of d^2 are:

X	Y	d	d^2
1	2	-1	1
2	1	1	1
3	3	0	0
4	5	-1	1
5	4	1	1
6	6	0	0
7	8	-1	1
8	9	-1	1
9	11	-2	4
10	10	0	0
11	12	-1	1
12	7	5	25
13	13	0	0
14	14	0	0
			36

$$\rho = 1 - \left[\frac{6\left(\sum d_i^2\right)}{n(n^2 - 1)} \right] = 0.921$$

$$\sigma_\rho = \sqrt{\frac{1}{(n-1)}} = 0.277$$

$$Z = \frac{\left(\rho - \mu_\rho\right)}{\sigma_\rho} = 3.325$$

Thus, the H_0 hypothesis is rejected at the 5% level ($Z = 3.325 > 1.96$). The rankings are highly (0.921) and significantly correlated.

6.7 GENERAL QUESTIONS AND ANSWERS

Question 1. Consider the following data for samples a and b drawn from non-normal populations:

a	5	6	9	2	5	3	2	7	8	
b	1	1	5	10	50	75	120	500	700	900

Test the claim that the samples come from identical populations.

Answer for #1:

The statistical software package results for the Wilcoxon Rank-Sum or Mann-Whitney Test are:

KRUSKAL-WALLIS ONE-WAY ANALYSIS
OF VARIANCE FOR 19 CASES

DEPENDENT VARIABLE IS   T

GROUPING VARIABLE IS   G$

GROUP	COUNT	RANK SUM
a	9	68.000
b	10	122.000

MANN-WHITNEY U TEST STATISTIC = 23.000

PROBABILITY IS   0.072

CHI-SQUARE APPROXIMATION = 3.244 WITH 1 DF

On the basis of the above results, the H_0, that the samples come from identical populations, cannot be rejected, and so is the claim.

Question 2. Consider the following data for samples m and n drawn from non-normal populations:

m	9	8	7	2	5	6	1	4	0
n	6	9	5	2	8	6	2	3	3

Test your claim that the samples do not come from identical populations.

Answer for #2:

The statistical software package results for the Sign Test are:

SIGN TEST RESULTS
COUNTS OF DIFFERENCES
(ROW VARIABLE GREATER THAN COLUMN)

	m	n
m	0	3
n	4	0

TWO-SIDED PROBABILITIES FOR EACH PAIR OF VARIABLES

	m	n
m	1.000	
n	1.000	1.000

On the basis of these results, the H_0 cannot be rejected. Therefore, your claim that the samples do not come from identical populations is not valid.

Question 3. Two candidates for office are about to lunch political campaigns. To determine whether x or y is more likeable, 1,000 potential voters were selected at random for testing. 490 voters reported that they liked x more than y, 35 found x and y equally good, and the remaining 475 found y better than x. Who is more likeable?

Answer for #3:

We are interested in testing the following hypotheses:

H_0: Voters are indifferent between x and y.
H_1: Voters are not indifferent between x and y.

The number of positive signs (x better than y) S = 490.

Thus,

n = 965 and

$$\mu_s = 0.5n = 482.5,$$

$$\sigma_s = \sqrt{0.25n} = 15.53, \text{ and}$$

$$Z = \frac{(S - \mu_s)}{\sigma_s} = 0.483$$

Therefore, at the 5% level Z < 1.96, and the H_0 hypothesis may not be rejected, which implies that the voters are indifferent between x and y.

Question 4. Consider the following data for samples k and r, drawn from non-normal populations:

k	74	88	54	12	46	89	12	47	45	63	12	18	14	16
r	85	82	84	26	34	24	21	78	45	64	21	11	15	48

Use the Wilcoxon Signed-Rank Test to test the claim that the samples come from identical populations.

Answer for #4:

Utilizing a statistical software package we get:

WILCOXON SIGNED RANKS TEST RESULTS

COUNTS OF DIFFERENCES
(ROW VARIABLE GREATER THAN COLUMN)

	k̲	r̲
k	0	4
r	9	0

$$Z = \frac{\text{(SUM OF SIGNED RANKS)}}{\text{SQUARE ROOT(SUM OF SQUARED RANKS)}}$$

	k̲	r̲
k	0.000	
r	1.224	0.000

TWO-SIDED PROBABILITIES USING NORMAL APPROXIMATION

	k̲	r̲
k	1.000	
r	0.221	1.000

The p-value of (0.221) indicates that the H_0 cannot be rejected and so is the claim.

Question 5. The manufacturer of a certain machine claims that all of his machines produce products of the same quality. To test his claim the buyer selected randomly four machines and let each machine produce 50 items. The number of non-defective items was used as a measure of quality, and the experiment was repeated 4 times. The number of non-defective items per run is reported as follows:

Run	Machine A	Machine B	Machine C	Machine D
1	49	44	49	38
2	50	47	44	33
3	48	49	47	43
4	50	50	44	42

The relevant hypotheses are:

H_0: The machines produce products of identical quality (or, the samples come from identical populations)
H_1: The machines do not produce products of identical quality (or, the samples do not come from identical populations)

Should we accept or reject the H_0? Why?

Answer for #5:

Utilizing a statistical software package we have:

KRUSKAL-WALLIS ONE-WAY ANALYSIS
OF VARIANCE FOR 16 CASES

DEPENDENT VARIABLE IS   M
GROUPING VARIABLE IS   M$

GROUP	COUNT	RANK SUM
A	4	52.000
B	4	41.500
C	4	32.500
D	4	10.000

KRUSKAL-WALLIS TEST STATISTIC = 10.778
PROBABILITY IS 0.013 ASSUMING CHI-SQUARE DISTRIBUTION WITH 3 DF

On the basis of the above results, the H_0 hypothesis is rejected at or less than the 5% level. Therefore, the machines do not produce products of identical qualities.

Question 6. Baseball teams a, b, c, d, e, f, g, h, i, j, k, l, m, and n, have been ranked by two different groups of judges as follows:

Team	a	b	c	d	e	f	g	h	i	j	k	l	m	n
Group X	1	2	3	4	5	6	7	8	9	10	11	12	13	14
Group Y	1	2	3	4	5	7	6	10	8	11	12	14	13	9

Does a significant rank correlation exist between X and Y?

Answer for #6:

The relevant hypotheses here are:

    H_0: The X and Y rankings are not correlated
    H_1: The X and Y rankings are correlated

$$\rho = 1 - \left[\frac{6\left(\sum d_i^2\right)}{n(n^2 - 1)} \right] = 0.9165$$

$$\sigma_\rho = \sqrt{\frac{1}{(n-1)}} = 0.277 \text{ and } Z = \frac{\left(\rho - \mu_\rho\right)}{\sigma_\rho} = 3.31$$

Thus, the H_0 hypothesis is rejected at the 5% level ($Z = 3.31 > 1.96$). The rankings are highly and significantly correlated.

6.8 HEALTH AND MEDICAL QUESTIONS AND ANSWERS

Question 1. Suppose we wish to compare the length of stay in the hospital for patients with the same diagnosis at two different hospitals. The following results are found:

Hospital 1	Hospital 2
21	86
10	27
32	10
60	68
8	87
44	76
29	125
5	60
13	35
26	73
33	96
	44
	238

a. Explain why a t-test may not be useful here.

b. Carry out a nonparametric procedure for testing the hypothesis that the lengths of stay are comparable in the two hospitals.

Answers for #1:

a. The distribution of length of stay is very skewed and far from being normal, which makes the t-test not very useful here.

b. Use the Wilcoxon rank-rum test (large-sample test).
Rank for Hospital 1 = 83.5, T = 3.10, p = 0.002.

Question 2. A common symptom of Otitis Media in young children is the pro-longed presence of fluid in the middle ear, known as *middle-ear effusion*. The presence of fluid may result in temporary hearing loss and interfere with normal learning skills in the first 2 years of life. One hypothesis is that babies whom are breast-fed for at least 1 month build up some immunity against the effects of the disease and have less prolonged effusion than do bottle-fed babies. A small study of 24 pairs of babies is set up, whereby the babies are matched on a one-to-one basis according to race, sex, socioeconomic status, and type of medications tak-en. One member of the matched pair is a breast-fed baby whereas the other member is a bottle-fed baby. The outcome variable is the duration of middle-ear effusion after the first episode of otitis media. Results are as follows:

Pair	Duration of Effusion in Breast-Fed Baby (days)	Duration of Effusion in Bottle-Fed Baby (days)
1	20	18
2	11	35
3	3	7
4	24	182
5	7	6
6	28	33
7	58	223
8	7	7
9	39	57
10	17	76
11	17	186
12	12	29
13	52	39
14	14	15
15	12	21
16	30	28
17	7	8
18	15	27
19	65	77
20	10	12
21	7	8
22	19	16
23	34	28
24	25	20

a. What hypotheses are being tested here?

b. Explain why a nonparametric test may be useful in testing these hypotheses.

c. Which nonparametric test should be used here?

d. Test the hypothesis that the duration of effusion is less prolonged among breast-fed babies than among bottle-fed babies using a nonparametric test.

Answer for #2:

a. H_0: $m_1 \geq m_2$
   H_1: $m_1 < m_2$

   Where:
   m_1 = median duration of effusion of breast-fed babies.
   m_2 = median duration of effusion of bottle-fed babies.

b. The distribution of duration of effusion is very skewed and far from being normal.

c. Wilcoxon signed-rank test

d. Rank $1 = 215$, $T = 2.38$, $p = 0.009$ (one-tail) $\Rightarrow$ breast-fed babies have a shorter duration of effusion than bottle-fed babies.

Appendix I:

PROJECTS

- **Project 1**
  AUTOMOBILE MILEAGE

- **Project 2**
  PHARMACOLOGICAL METHODOLOGY

- **Project 3**
  CHUCK-A-LUCK GAME

- **Project 4**
  SAMPLE TESTING

- **Project 5**
  COLLEGE STUDENT SURVEY

∧∧

PROJECT 1: AUTOMOBILE MILEAGE

Table 1 contains mileage (M) and weight (W) sample data for various models of domestic (D) and foreign (F) cars.

(a) For M, report: size, maximum and minimum; compute raw: mean, variance, standard deviation, median, 1st and 3rd quartiles, mode or modes, skewness, kurtosis, coefficient of variation and standard error.

n = 74.000

MAXIMUM = 41.000

MINIMUM = 12.000

MEAN = 21.297

VARIANCE = 33.472

STANDARD DEV = 5.786

MEDIAN = 20.000

1st QUARTILE = 18

2nd QUARTILE = 25

MODES = 18

SKEWNESS = 0.949

KURTOSIS = 0.975

C.V. = 0.272

STD. ERROR = 0.673

∧∧

^^

(b) Which cars rank top three in terms of mileage? Which cars rank worst three in terms of mileage?

Consider Table 2 which shows the data ranked from low to high.

Thus,

Top three: 41: "VW Rabbit Ds"
                35: "Subaru," "Datsun 210"
                34: "Plymouth Champ"

Worst three: 12: "Lincoln Continental," "Lincoln Mark V"
                14: "Cadillac Deville," "Cadillac Eldorado," "Lincoln Ver-
                sails," "Mercury Cougar," "Mercury Cougar RX," "Peugeot
                604"
                15: "Buick Electra" (15), "Mercury Marquis"

^^

(c) Draw the box-and-whiskers plot for M given D; in the same diagram, draw the box-and-whiskers plot for M given F.

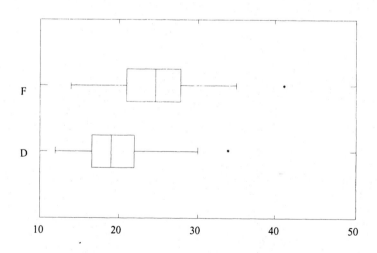

^^

∧∧

(d)   Draw the stem-and leaf plot for M.

1		22
1		44444455
1		66667777
1	H	888888888999999999
2	M	00011111
2		22222333
2	H	444455555
2		666
2		8889
3		001
3		
3		455
		OUTSIDE VALUES
4		1

∧∧

^^

(e) Tabulate the ungrouped probability distributions for M and W.

MILEAGE (M)

COUNT	PERCENT	M
2	2.7	12.000
6	8.1	14.000
2	2.7	15.000
4	5.4	16.000
4	5.4	17.000
9	12.2	18.000
8	10.8	19.000
3	4.1	20.000
5	6.8	21.000
5	6.8	22.000
3	4.1	23.000
4	5.4	24.000
5	6.8	25.000
3	4.1	26.000
3	4.1	28.000
1	1.4	29.000
2	2.7	30.000
1	1.4	31.000
1	1.4	34.000
2	2.7	35.000
1	1.4	41.000

^^

∧∧∧

(f) For M and W draw: ungrouped histograms, scatter plot and regression line.

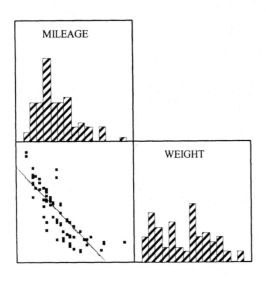

∧∧∧

(g) For M, tabulate the grouped probability distribution (5 intervals), graph the histogram and compute the grouped mean, variance and standard deviation.

Intervals	Frequency (f)	Probability	Midpoint (m)
12.0-17.8	18	18/74	14.9
17.9-23.6	33	33/74	20.75
23.7-29.4	16	16/74	26.55
29.5-35.2	6	6/74	32.35
35.3-41.0	1	1/74	38.15

Probability

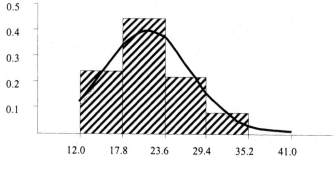

Grouped Mean = 21.757

Grouped Variance = 29.991

Grouped Standard Deviation = 5.476

∧∧

(h) For M, compute the 95% confidence interval about the population mean and variance.

μ = Sample Average ± (1.96)(Standard Error) = 21.297 ± 1.32 with 95% Confidence.

and

$P\{ [(n-1)(s^2) / \chi^2_L] = \sigma^2 = [(n-1)(s^2) / \chi^2_U] \} = 0.95$
$P\{ [(73)(33.472) / 117.76] = \sigma^2 = [(73)(33.472) / 66.03] \} = 0.95$
$P (20.75 = \sigma^2 = 37.01) = 0.95.$

∧∧

.

∧∧∧

(i) For M, test your claim, at the 95% level, that the population mean is less than or equal to 20 and the population variance is greater than or equal to 32.

Test for μ: H_0: $\mu \leq 20$
                           H_1: $\mu > 20$
                           $Z_\alpha = 1.645$, $Z_{st} = 1.93$. Because the $Z_{st} > Z_\alpha$, the H_0 is rejected.

Test for σ^2: H_0: $\sigma^2 \geq 32$
                           H_1: $\sigma^2 < 32$
                           $\chi_\alpha^2 = 112.825$, $\chi_{st}^2 = 76.358$. Because the $\chi_{st}^2 < \chi_\alpha^2$, the H_0 is rejected.

∧∧

(j) Compute the correlation coefficient between M and W and discuss its statistical significance.

$r_{MW} = -0.823$ (P-value $= 0.000$).

The correlation coefficient is statistically significant.

∧∧

(k) Compute a linear regression of M against W and discuss its statistical significance and usage.

$M = 39.583 - 0.006W$	t Stats:	(25.755) (-12.289)
$R^2 = 0.677$	P-values:	(0.10E-14) (0.10E-14)

As one should expect, W affects M negatively. The value of the R^2 indicates that only 67.7% of M's variation is explained by W. The t Statistics and the P-values indicate that the estimated coefficients are statistically significant.

The estimated regression may be used to show how, ceteris paribus, changes in W affect average M. For example, at $W = 2000$, average $M \sim 28$; when the weight decreases to $W = 1,000$, average mileage improves to $M \sim 34$. Hence, ceteris paribus, a 1000 decrease in weight improves average mileage by about 6 miles per gallon.

∧∧

TABLE 1 (CONTINUED ON NEXT PAGE)

CAR	MILEAGE	WEIGHT	ORIGIN
"AMC Concord"	22.000	2,930.000	" D"
"AMC Pacer"	17.000	3,350.000	" D"
"AMC Spirit"	22.000	2,640.000	" D"
"Audi 5000"	17.000	2,830.000	" F"
"Audi Fox"	23.000	2,070.000	" F"
"BMW 320i"	25.000	2,650.000	" F"
"Buick Century"	20.000	3,250.000	" D"
"Buick Electra"	15.000	4,080.000	" D"
"Buick Le Sabre"	18.000	3,670.000	" D"
"Buick Opel"	26.000	2,230.000	" D"
"Buick Regal"	20.000	3,280.000	" D"
"Buick Riviera"	16.000	3,880.000	" D"
"Buick Skylark"	19.000	3,400.000	" D"
"Cadillac Deville"	14.000	4,330.000	" D"
"Cadillac Eldorado"	14.000	3,900.000	" D"
"Cadillac Seville"	21.000	4,290.000	" D"
"Chevy Chevette"	29.000	2,110.000	" D"
"Chevy Impala"	16.000	3,690.000	" D"
"Chevy Malibu"	22.000	3,180.000	" D"
"Chevy Monte Carlo"	22.000	3,220.000	" D"
"Chevy Monza"	24.000	2,750.000	" D"
"Chevy Nova"	19.000	3,430.000	" D"
"Datsun 200SX"	23.000	2,370.000	" F"
"Datsun 210"	35.000	2,020.000	" F"
"Datsun 510"	24.000	2,280.000	" F"
"Datsun 810"	21.000	2,750.000	" F"
"Dodge Colt"	30.000	2,120.000	" D"
"Dodge Diplomat"	18.000	3,600.000	" D"
"Dodge Magnum"	16.000	3,870.000	" D"
"Dodge St Regis"	17.000	3,740.000	" D"
"Fiat Strada"	21.000	2,130.000	" F"
"Ford Fiesta"	28.000	1,800.000	" D"
"Ford Mustang"	21.000	2,650.000	" D"
"Honda Accord"	25.000	2,240.000	" F"
"Honda Civic"	28.000	1,760.000	" F"
"Lincoln Continental"	12.000	4,840.000	" D"
"Lincoln Mark V"	12.000	4,720.000	" D"
"Lincoln Versails"	14.000	3,830.000	" D"
"Mazda GLC"	30.000	1,980.000	" F"
"Mercury Bobcat"	22.000	2,580.000	" D"
"Mercury Cougar"	14.000	4,060.000	" D"
"Mercury Cougar RX"	14.000	4,130.000	" D"
"Mercury Marquis"	15.000	3,720.000	" D"
"Mercury Monarch"	18.000	3,370.000	" D"
"Mercury Zephyr"	20.000	2,830.000	" D"
"Oldsmobile 98"	21.000	4,060.000	" D"
"Oldsmobile Cutlass"	19.000	3,300.000	" D"
"Oldsmobile Cutlass Supreme"	19.000	3,310.000	" D"
"Oldsmobile Delta88"	18.000	3,690.000	" D"

"Oldsmobile Omega"	19.000	3,370.000	"	D"
"Oldsmobile Starfire"	24.000	2,720.000	"	D"
"Olds mobile Toranado"	16.000	4,030.000	"	D"
"Peugeot 604"	14.000	3,420.000	"	F"
"Plymouth Arrow"	28.000	2,360.000	"	D"
"Plymouth Champ"	34.000	1,800.000	"	D"
"Plymouth Horizon"	25.000	2,200.000	"	D"
"Plymouth Sapporo"	26.000	2,520.000	"	D"
"Plymouth Volare"	18.000	3,330.000	"	D"
"Pontiac Catalina"	18.000	3,700.000	"	D"
"Pontiac Firebird"	18.000	3,470.000	"	D"
"Pontiac GrandPrix"	19.000	3,210.000	"	D"
"Pontiac Le Mans"	19.000	3,200.000	"	D"
"Pontiac Phoenix"	19.000	3,420.000	"	D"
"Pontiac Sunbird"	24.000	2,690.000	"	D"
"Renault Le Car"	26.000	1,830.000	"	F"
"Subaru"	35.000	2,050.000	"	F"
"Toyota Celica"	18.000	2,410.000	"	F"
"Toyota Corolla"	31.000	2,200.000	"	F"
"Toyota Corona"	18.000	2,670.000	"	F"
"VW Rabbit"	25.000	1,930.000	"	F"
"VW Rabbit Ds"	41.000	2,040.000	"	F"
"VW Scirocco"	25.000	1,990.000	"	F"
"VW Dasher"	23.000	2,160.000	"	F"
"Volvo 260"	17.000	3,170.000	"	F"

TABLE 2 (CONTINUED ON NEXT PAGE)

CAR	MILEAGE	WEIGHT	ORIGIN	
"Lincoln Continental"	12.000	4,840.000	"	D"
"Lincoln Mark V"	12.000	4,720.000	"	D"
"Cadillac Deville"	14.000	4,330.000	"	D"
"Cadillac Eldorado"	14.000	3,900.000	"	D"
"Lincoln Versails"	14.000	3,830.000	"	D"
"Mercury Cougar"	14.000	4,060.000	"	D"
"Mercury Cougar RX"	14.000	4,130.000	"	D"
"Peugeot 604"	14.000	3,420.000	"	F"
"Buick Electra"	15.000	4,080.000	"	D"
"Mercury Marquis"	15.000	3,720.000	"	D"
"Buick Riviera"	16.000	3,880.000	"	D"
"Chevy Impala"	16.000	3,690.000	"	D"
"Dodge Magnum"	16.000	3,870.000	"	D"
"Oldsmobile Toranado"	16.000	4,030.000	"	D"
"AMC Pacer"	17.000	3,350.000	"	D"
"Audi 5000"	17.000	2,830.000	"	F"
"Dodge St Regis"	17.000	3,740.000	"	D"
"Volvo 260"	17.000	3,170.000	"	F"
"Buick Le Sabre"	18.000	3,670.000	"	D"
"Dodge Diplomat"	18.000	3,600.000	"	D"
"Mercury Monarch"	18.000	3,370.000	"	D"
"Oldsmobile Delta88"	18.000	3,690.000	"	D"
"Plymouth Volare"	18.000	3,330.000	"	D"
"Pontiac Catalina"	18.000	3,700.000	"	D"
"Pontiac Firebird"	18.000	3,470.000	"	D"
"Toyota Celica"	18.000	2,410.000	"	F"
"Toyota Corona"	18.000	2,670.000	"	F"
"Buick Skylark"	19.000	3,400.000	"	D"
"Chevy Nova"	19.000	3,430.000	"	D"
"Oldsmobile Cutlass Supreme"	19.000	3,310.000	"	D"
"Oldsmobile Cutlass"	19.000	3,300.000	"	D"
"Oldsmobile Omega"	19.000	3,370.000	"	D"
"Pontiac GrandPrix"	19.000	3,210.000	"	D"
"Pontiac Le Mans"	19.000	3,200.000	"	D"
"Pontiac Phoenix"	19.000	3,420.000	"	D"
"Buick Century"	20.000	3,250.000	"	D"
"Buick Regal"	20.000	3,280.000	"	D"
"Mercury Zephyr"	20.000	2,830.000	"	D"
"Cadillac Seville"	21.000	4,290.000	"	D"
"Datsun 810"	21.000	2,750.000	"	F"
"Fiat Strada"	21.000	2,130.000	"	F"
"Ford Mustang"	21.000	2,650.000	"	D"
"Oldsmobile 98"	21.000	4,060.000	"	D"
"AMC Concord"	22.000	2,930.000	"	D"
"AMC Spirit"	22.000	2,640.000	"	D"
"Chevy Malibu"	22.000	3,180.000	"	D"
"Chevy Monte Carlo"	22.000	3,220.000	"	D"
"Mercury Bobcat"	22.000	2,580.000	"	D"
"Audi Fox"	23.000	2,070.000	"	F"

"Datsun 200SX"	23.000	2,370.000	"	F"
"VW Dasher"	23.000	2,160.000	"	F"
"Chevy Monza"	24.000	2,750.000	"	D"
"Datsun 510"	24.000	2,280.000	"	F"
"Oldsmobile Starfire"	24.000	2,720.000	"	D"
"Pontiac Sunbird"	24.000	2,690.000	"	D"
"BMW 320i"	25.000	2,650.000	"	F"
"Honda Accord"	25.000	2,240.000	"	F"
"Plymouth Horizon"	25.000	2,200.000	"	D"
"VW Rabbit"	25.000	1,930.000	"	F"
"VW Scirocco"	25.000	1,990.000	"	F"
"Buick Opel"	26.000	2,230.000	"	D"
"Plymouth Sapporo"	26.000	2,520.000	"	D"
"Renault Le Car"	26.000	1,830.000	"	F"
"Ford Fiesta"	28.000	1,800.000	"	D"
"Honda Civic"	28.000	1,760.000	"	F"
"Plymouth Arrow"	28.000	2,360.000	"	D"
"Chevy Chevette"	29.000	2,110.000	"	D"
"Dodge Colt"	30.000	2,120.000	"	D"
"Mazda GLC"	30.000	1,980.000	"	F"
"Toyota Corolla"	31.000	2,200.000	"	F"
"Plymouth Champ"	34.000	1,800.000	"	D"
"Datsun 210"	35.000	2,020.000	"	F"
"Subaru"	35.000	2,050.000	"	F"
"VW Rabbit Ds"	41.000	2,040.000	"	F"

∧∧

PROJECT 2: PHARMACOLOGICAL METHODOLOGY

A pharmaceutical company has modified a new methodology for diagnosing breast cancer. The previous method had a success rate of about 45%. To test the new procedure, the company contacted a screening test on 5000 individuals with the following results:

	Cancer Present	Cancer Absent	Total
Positive Test	2,570	1,850	4,420
Negative Test	300	280	580
Total	2,870	2,130	5,000

(a) Compute: Sensitivity, Specificity, Positive Predictive Value, Negative Predictive Value and the Posterior Odds. Interpret the meaning of Posterior Odds.

Sensitivity = 2,570 / 2,870 = 0.90
Specificity = 280 / 2,130 = 0.13
Positive Predictive Value = 2,570 / 4,420 = 0.58
Negative Predictive Value = 280 / 580 = 0.48
Posterior Odds = $(0.45)[(0.90) / (1-0.13)] = 0.47$

∧∧

(b) Compute the 95% confidence interval about the population sensitivity and specificity.

$s_{Sen} = 0.0042$,
$Sensitivity_{Population} = 0.90 \pm (1.96)(0.0042) = 0.90 \pm 0.0082$
with 95% Confidence.

$s_{Spe} = 0.0048$,
$Sensitivity_{Population} = 0.13 \pm (1.96)(0.0048) = 0.13 \pm 0.0093$
with 95% Confidence.

∧∧

∧∧

(c) Given that cancer is present, test, at the 5% level, the claim that the probability of a positive test (P_1) is equal to the probability of a negative test (P_2).

$H_0 : P_1 = P_2$
$H_1 : P_1 \neq P_2$

$\chi^2_{st} = 8.38$, df $= 1$. Since $\chi^2_{st} = 8.38 > \chi^2_{critical} = 3.84$, the H_0 is rejected.

∧∧

^^

PROJECT 3: CHUCK-A-LUCK GAME

Consider the chuck-a-luck game (X) with dice i and j. I am the house, you are the player, and I plan to roll the dice once:

(a) If your favorite number appears twice, I will pay you $R_1 = \$5$. If your favorite number appears once, I will pay you $R_2 = \$2$. To play the game, you have to pay me a fee of $F = \$1.5$. Compute the game's expected value and risk. Is the game fair? Is it more than fair to the house or the player? What fee (F) will make the game fair? Draw the game's probability distribution or bar chart.

$E(X) = -0.806$, Risk $= 0.562$; the game is more than fair to the house.

The fee that will make the game fair is $F = 0.694$

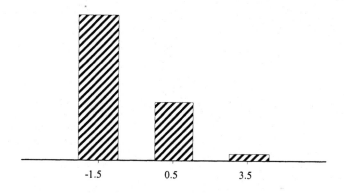

| | | |
| -1.5 | 0.5 | 3.5 |

^^

∧∧∧

(b) Redo (a) with $R_1 = \$6$.

$E(X) = -0.778$, Risk $= 0.768$; the game is more than fair to the house.

The fee that will make the game fair is $F = 0.722$

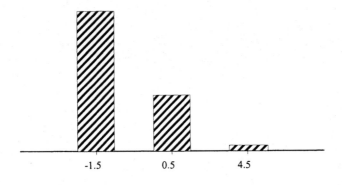

-1.5	0.5	4.5

∧∧∧

(c) Redo (a) with $F=0.5$

$E(X) = 0.194$, Risk $= 1.054$; the game is more than fair to the player.

The fee that will make the game fair is $F = 0.722$

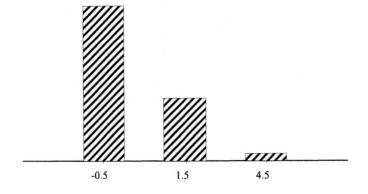

-0.5	1.5	4.5

∧∧∧

^^^

PROJECT 4: SAMPLE TESTING

(1) Consider the information for a sample in the following table:

	Test:
Size = 100	
	$H_0 \geq 120$
Average = 115	$H_1 < 120$
Standard Deviation = 25	$\alpha = 0.05$

(a) Should the H_0 be rejected? Why?

The H_0 should be rejected because $Z_\alpha = -2 < Z_{st} = -1.645$.

^^

(b) Let $\mu_\tau = 118$. Compute the Power of the test and interpret your findings.

$\beta = P[(Z = \{(120-118)/(25/10) - 1.645\}]$

$= P(Z = -0.845) = 0.7995$

Power $= 0.2005$

Thus, there is a 20.05% chance of rejecting the H_0 given that H_0 is false and the true population mean is 118.

^^

∧∧

(c) Let alternative μ_τ values be: 120, 119, 118, 117, 116, 115, 114, 113, 112, and 111. Compute the β and Power values and graph the corresponding Power curve.

μ_τ	β	Power
120	0.950	0.050
119	0.894	0.106
118	0.799	0.201
117	0.670	0.330
116	0.516	0.484
115	0.363	0.637
114	0.224	0.776
113	0.123	0.877
112	0.059	0.941
111	0.025	0.975

Power

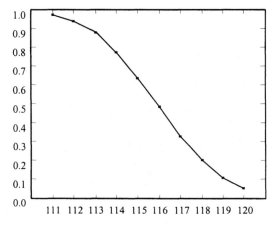

∧∧

^^

(2) Consider the information for a sample in the following table:

	Test:
Size = 25	
	$H_0 \leq 180$
Average = 200	$H_1 > 180$
Standard Deviation = 46	$\alpha = 0.05$

(a) Should the H_0 be rejected? Why?

The H_0 should be rejected because $Z_\alpha = 2.17 > Z_{st} = 1.645$.

^^

(b) Let $\mu_\tau = 190$. Compute the Power of the test and interpret your findings.

$\beta = P[(Z = \{(180\text{-}190)/(46/5) + 1.645\}]$

$= P(Z = 0.558) = 0.7088$

Power $= 0.2912$

Thus, there is a 29.12% chance of rejecting the H_0 given that H_0 is false and the true population mean is 190.

^^

(c) Let alternative μ_τ values be: 180, 190, 200, 210, 220, and 230.
Compute the β and Power values and graph the corresponding Power curve.

μ_τ	β	Power
180	0.950	0.050
190	0.709	0.291
200	0.298	0.702
210	0.053	0.947
220	0.004	0.997
230	0.000	1.000

Power

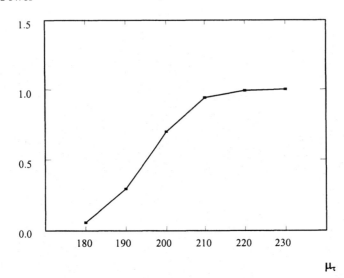

μ_τ

∧∧

(3) Consider the information for a sample in the following table:

	Test:
Size = 64	
	$H_0 = 70$
Average = 80	$H_1 \neq 70$
Standard Deviation = 64	$\alpha = 0.05$

(a) Should the H_0 be rejected? Why?

The H_0 should be rejected because $Z_{\alpha/2} = -1.96 > Z_{st} = -2.67$.

∧∧

^^^

(b) Let $\mu_\tau = 75$. Compute the Power of the test and interpret your findings.

$$\beta = P[\ (70-75)/(30/8) - 1.96 = Z = (70-75)/(30/8) + 1.96\ ] = 0.736$$

Power $= 0.264$

Thus, there is a 26.4% chance of rejecting the H_0 given that H_0 is false and the true population mean is 74.

^^^

(c) Let alternative μ_τ values be: 40, 45, 50, 55, 60, 65, 70, 75, 80, 85, 90, 95, 100. Compute the β and Power values and graph the corresponding Power curve.

μ_τ	β	Power
40	0.000	1.000
45	0.000	1.000
50	0.000	1.000
55	0.021	0.979
60	0.239	0.761
65	0.736	0.264
70	0.950	0.050
75	0.736	0.264
80	0.239	0.761
85	0.021	0.979
90	0.000	1.000
95	0.000	1.000
100	0.000	1.000

Power

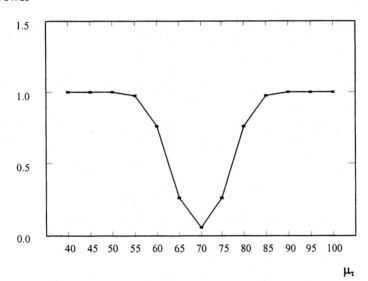

^^

PROJECT 5: COLLEGE STUDENT SURVEY

The data in the following tables originated from a pilot sample of 60 female and 60 male college students. The students were asked the following questions and their answers were recorded in the following tables.

Questions:

SEX: What is your sex?
SAT: What is your average SAT?
GPA: What is your GPA?
Y: What is your parents' income?
EX: How many hours do you exercise per day?
JUNK: How much do you spend on junk food per day?
P: What is your political affiliation?

Notes:
S = severity of exercise: A ≤ 1, 1 < B < 2, E ≥ 2.
The Y numbers have been divided by 1,000.
For P: D = Democratic, R = Republican, I = Independent, O = Other

^^

MALES

1	2	3	4	5	6	7
SAT	GPA	Y	EX	S	JUNK	P
510	3.0	80	2.0	E	2	R
540	3.2	90	3.0	E	1	I
510	2.5	100	1.0	A	2	O
540	3.2	70	1.0	A	3	D
500	3.0	60	1.5	B	1	D
570	2.8	70	1.5	B	2	D
540	3.0	65	1.0	A	1	D
520	3.1	60	1.5	B	0	R
530	1.8	65	2.0	E	2	D
510	2.0	80	4.0	E	1	R
570	2.5	120	3.0	E	2	D
560	3.0	60	1.0	A	2	R
540	2.7	80	2.0	E	2	D
580	2.8	120	3.0	E	2	D
610	3.2	150	1.5	B	2	D
520	2.5	200	3.0	E	2	D
510	2.4	40	1.0	A	2	D
550	2.0	60	2.0	E	2	R
560	2.7	70	2.0	E	1	R
580	2.9	80	2.0	E	3	R
540	3.1	50	2.0	E	1	D
520	2.5	48	1.5	B	2	D
510	2.1	50	1.0	A	3	O
590	2.8	70	1.0	A	2	D
610	3.3	80	1.0	A	1	D
650	3.5	90	1.0	A	1	R
660	3.0	170	1.0	A	1	R
580	2.9	85	1.5	B	1	I
670	3.8	130	1.0	A	1	I
640	3.6	160	1.0	A	0	O
590	3.0	80	1.5	B	2	O
600	3.1	140	1.0	A	3	O
530	2.6	60	3.0	E	2	O
520	2.0	80	2.0	E	1	I
510	1.8	70	2.0	E	3	I
680	3.9	60	1.0	A	2	D
640	3.7	90	1.5	B	3	D
505	2.1	80	3.0	E	2	R
570	2.8	78	1.0	A	2	R
605	3.1	85	1.5	B	2	R
610	3.0	130	1.0	A	4	R
585	2.7	100	1.0	A	3	R
640	3.6	150	1.5	B	2	D
520	1.7	110	2.0	E	4	R
500	1.9	70	3.0	E	2	D
590	2.0	80	1.5	B	3	R
620	3.5	140	1.0	A	2	D
540	2.8	85	2.0	E	3	R
530	2.3	74	1.5	B	2	D
510	2.0	70	2.0	E	3	O
500	1.4	65	1.5	B	3	I
560	3.0	70	1.0	A	3	I
570	3.1	78	1.0	A	3	I
640	3.9	200	1.5	B	2	I
680	4.0	500	1.0	A	2	I
690	4.0	450	1.0	A	2	R
740	3.9	300	0.5	A	2	R
780	4.0	200	1.0	A	2	R
570	3.0	90	1.5	B	2	D
520	2.5	80	2.0	E	2	R

FEMALES

8	9	10	11	12	13	14
SAT	GPA	Y	EX	S	JUNK	P
590	2.9	100	1.5	B	2	D
580	3.0	150	2.0	E	1	D
630	3.9	180	1.0	A	1	D
620	3.8	170	1.5	B	2	D
570	3.1	150	2.0	E	1	R
580	3.2	155	1.0	A	1	R
510	2.6	100	2.0	E	0	R
530	2.9	120	1.0	A	0	D
590	3.0	160	1.0	A	0	D
620	3.5	200	1.5	B	1	D
650	3.6	250	1.0	A	2	D
690	3.9	300	1.0	A	1	I
580	3.7	120	1.5	B	1	I
570	3.0	70	1.5	B	1	I
540	2.7	85	1.0	A	1	I
530	2.2	90	2.0	E	1	I
510	2.0	80	3.0	E	1	O
600	2.9	110	1.0	A	1	O
680	3.8	150	1.0	A	2	O
630	3.4	140	0.5	A	2	I
520	2.8	120	2.0	E	1	D
505	2.1	85	2.0	E	1	R
550	2.7	110	1.5	B	2	D
550	2.2	80	1.0	A	3	R
570	2.6	90	1.5	B	2	D
605	3.1	150	1.0	A	3	R
600	3.7	160	1.0	A	2	D
670	3.9	280	1.0	A	2	D
660	3.8	270	1.0	A	1	D
580	2.5	95	0.5	A	1	D
540	2.5	85	2.0	E	1	D
550	2.4	70	2.0	E	1	D
520	2.3	68	2.0	E	1	D
540	2.4	70	1.5	B	2	I
580	2.6	78	0.5	A	2	O
720	4.0	90	0.5	A	2	I
730	4.0	120	0.5	A	3	O
690	3.9	89	0.5	A	2	R
630	3.1	110	1.0	A	2	D
640	3.2	140	0.5	A	2	R
620	3.1	100	0.5	A	1	I
610	3.0	120	0.5	A	2	D
590	2.5	95	1.0	A	3	D
570	2.4	100	1.0	A	2	D
680	4.0	150	1.0	A	2	D
630	3.8	140	1.0	A	1	D
650	3.5	145	1.0	A	2	I
660	3.6	160	1.0	A	1	I
580	2.6	85	2.0	E	1	I
590	2.7	95	2.0	E	1	D
570	2.8	57	2.0	E	1	I
560	2.6	80	1.5	B	1	D
530	2.3	70	1.0	A	1	I
520	2.2	60	3.0	E	1	I
560	2.6	90	1.0	A	2	D
670	3.7	100	1.0	A	2	R
620	3.2	120	1.0	A	1	D
610	3.0	110	1.0	A	1	D
600	3.5	67	0.5	A	1	I
580	3.0	87	1.0	A	1	O

^^^

(a) Display of descriptive statistics for all students in the sample.

DESCRIPTIVE STATISTICS FOR ALL STUDENTS:

	SAT	GPA	Y	EX	JUNK
N	120	120	120	120	120
LO 95% CI	574.31	2.8408	102.32	1.3137	1.5769
MEAN	585.13	2.9525	114.91	1.4375	1.7250
UP 95% CI	595.94	3.0642	127.50	1.5613	1.8731
SD	59.837	0.6178	69.650	0.6847	0.8195
VARIANCE	3580.4	0.3817	4851.2	0.4688	0.6716
SE MEAN	5.4623	0.0564	6.3582	0.0625	0.0748
C.V.	10.226	20.925	60.614	47.628	47.509
MINIMUM	500.00	1.4000	40.000	0.5000	0.0000
1ST QUARTI	540.00	2.5000	75.000	1.0000	1.0000
MEDIAN	580.00	3.0000	90.000	1.0000	2.0000
3RD QUARTI	620.00	3.5000	140.00	2.0000	2.0000
MAXIMUM	780.00	4.0000	500.00	4.0000	4.0000
SKEW	0.6934	-0.0433	2.9113	1.1139	0.2673
KURTOSIS	0.0674	-0.6832	11.100	1.2266	-0.0841

^^^

(a1) (i) Grouped Histograms for SAT, GPA, Y, EX and JUNK. (ii) Box and Whiskers Plots for SAT, GPA, Y, EX and JUNK by SEX. (iii) Stem and Leaf Plots for SAT and GPA

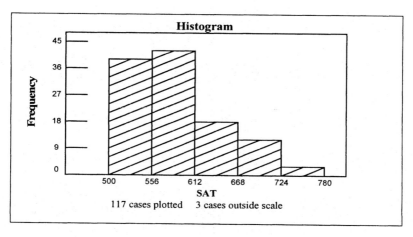

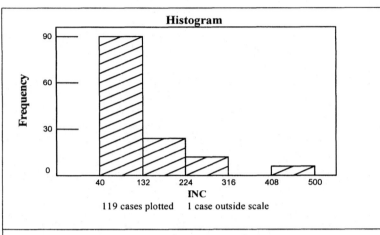

Histogram

119 cases plotted   1 case outside scale

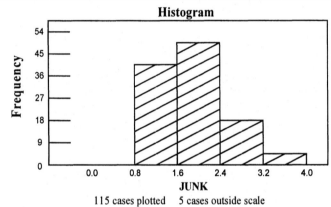

Histogram

115 cases plotted   5 cases outside scale

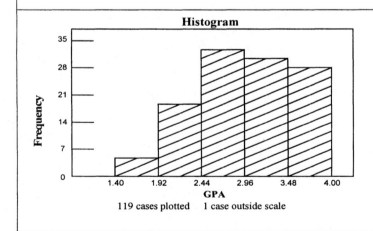

Histogram

119 cases plotted   1 case outside scale

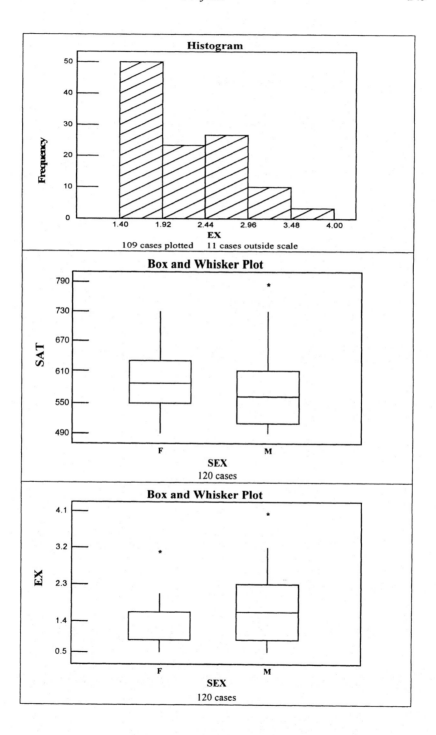

Histogram

109 cases plotted   11 cases outside scale

Box and Whisker Plot

120 cases

Box and Whisker Plot

120 cases

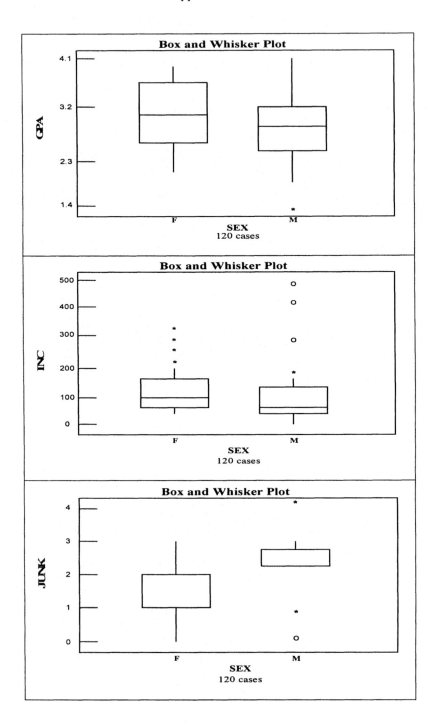

^^

STEM AND LEAF PLOT OF SAT

```
LEAF DIGIT UNIT = 10              MINIMUM  500.00
5  0  REPRESENTS 500.             MEDIAN   580.00
                                  MAXIMUM  780.00

      STEM  LEAVES
       14   5  00000111111111
       29   5  222222222333333
       42   5  4444444445555
       57   5  666667777777777
      (18)  5  888888888889999999
       45   6  00000011111
       34   6  222223333
       25   6  44444555
       17   6  666777
       11   6  8888999
        4   7
        4   7  23
        2   7  4
        1   7
        1   7  8
```

STEM AND LEAF PLOT OF GPA

```
LEAF DIGIT UNIT = 0.1             MINIMUM  1.4000
1  4  REPRESENTS 1.4              MEDIAN   3.0000
                                  MAXIMUM  4.0000

STEM  LEAVES
    1   1  4
    2   1  7
    5   1  889
   14   2  000000111
   20   2  222333
   32   2  444455555555
   45   2  6666666777777
   57   2  888888899999
  (24)  3  000000000000000111111111
   39   3  2222223
   32   3  455555
   26   3  66667777
   18   3  888889999999
    6   4  000000
```

^^

^^

(a2) Display of GPA averages by PA, SEX and S.

(S.D. = Standard Deviation, S.E. = Standard Error.)

BREAKDOWN FOR GPA

VARIABLE	LEVEL	N	MEAN	S.D.	S.E.
PA	D	17	3.2294	0.5554	0.1347
PA	I	9	3.3333	0.5500	0.1833
PA	O	5	3.2600	0.6066	0.2713
PA	R	6	3.2167	0.5913	0.2414
SEX	F	37	3.2568	0.5439	0.0894
PA	D	7	3.1571	0.4860	0.1837
PA	I	4	3.4750	0.4992	0.2496
PA	O	4	2.8250	0.6602	0.3301
PA	R	9	3.3222	0.5310	0.1770
SEX	M	24	3.2167	0.5411	0.1104
S	A	61	3.2410	0.5386	0.0690
PA	D	6	3.0167	0.5115	0.2088
PA	I	3	3.0333	0.6506	0.3756
SEX	F	9	3.0222	0.5191	0.1730
PA	D	8	3.0125	0.4883	0.1726
PA	I	3	2.7333	1.2583	0.7265
PA	O	1	3.0000		
PA	R	3	2.7333	0.6351	0.3667
SEX	M	15	2.9000	0.6503	0.1679
S	B	24	2.9458	0.5956	0.1216
PA	D	6	2.6167	0.2639	0.1078
PA	I	4	2.4500	0.3000	0.1500
PA	O	1	2.0000		
PA	R	3	2.6000	0.5000	0.2887
SEX	F	14	2.5214	0.3378	0.0903
PA	D	7	2.4714	0.4716	0.1782
PA	I	3	2.3333	0.7572	0.4372
PA	O	2	2.3000	0.4243	0.3000
PA	R	9	2.4111	0.4702	0.1567
SEX	M	21	2.4095	0.4742	0.1035
S	E	35	2.4543	0.4231	0.0715
OVERALL		120	2.9525	0.6178	0.0564

CASES INCLUDED 120   MISSING CASES 0

^^

∧∧

(a3) Display of SAT averages by PA, SEX and S.

BREAKDOWN FOR SAT

VARIABLE	LEVEL	N	MEAN	S.D.	S.E.
PA	D	17	612.35	40.393	9.7968
PA	I	9	626.67	63.246	21.082
PA	O	5	634.00	67.676	30.265
PA	R	6	622.50	53.828	21.975
SEX	F	37	620.41	50.914	8.3701
PA	D	7	584.29	58.554	22.131
PA	I	4	620.00	63.770	31.885
PA	O	4	565.00	65.574	32.787
PA	R	9	649.44	76.667	25.556
SEX	M	24	611.46	72.163	14.730
S	A	61	616.89	59.758	7.6512
PA	D	6	585.00	30.166	12.315
PA	I	3	563.33	20.817	12.019
SEX	F	9	577.78	28.186	9.3953
PA	D	8	572.50	53.918	19.063
PA	I	3	573.33	70.238	40.552
PA	O	1	590.00		
PA	R	3	571.67	45.369	26.194
SEX	M	15	573.67	49.730	12.840
S	B	24	575.21	42.259	8.6261
PA	D	6	550.00	29.665	12.111
PA	I	4	550.00	29.439	14.720
PA	O	1	510.00		
PA	R	3	528.33	36.171	20.883
SEX	F	14	542.50	30.176	8.0648
PA	D	7	540.00	27.689	10.465
PA	I	3	523.33	15.275	8.8192
PA	O	2	520.00	14.142	10.000
PA	R	9	532.78	26.114	8.7047
SEX	M	21	532.62	24.167	5.2737
S	E	35	536.57	26.755	4.5225
OVERALL		120	585.12	59.837	5.4623

CASES INCLUDED 120   MISSING CASES 0

∧∧

^^^

(a4) Display of descriptive statistics for female college students.

DESCRIPTIVE STATISTICS
FOR FEMALE COLLEGE STUDENTS:

	SAT	GPA	Y	EX	JUNK
N	60	60	60	60	60
LO 95% CI	581.70	2.9004	107.45	1.1052	1.2531
MEAN	595.83	3.0500	121.18	1.2583	1.4333
UP 95% CI	609.97	3.1996	134.91	1.4115	1.6136
SD	54.712	0.5792	53.145	0.5930	0.6979
VARIANCE	2993.4	0.3354	2824.4	0.3516	0.4870
SE MEAN	7.0632	0.0748	6.8609	0.0766	0.0901
C.V.	9.1824	18.989	43.855	47.124	48.688
MINIMUM	505.00	2.0000	57.000	0.5000	0.0000
1ST QUARTI	552.50	2.6000	85.000	1.0000	1.0000
MEDIAN	590.00	3.0000	105.00	1.0000	1.0000
3RD QUARTI	630.00	3.6000	150.00	1.5000	2.0000
MAXIMUM	730.00	4.0000	300.00	3.0000	3.0000
SKEW	0.4140	0.1283	1.6185	0.8823	0.3905
KURTOSIS	-0.4564	-1.1624	2.6086	0.5538	-0.0892

^^^

(a5) Display of descriptive statistics for male college students.

DESCRIPTIVE STATISTICS FOR MALE COLLEGE STUDENTS:

	SAT	GPA	Y	EX	JUNK
N	60	60	60	60	60
LO 95% CI	558.09	2.6886	87.206	1.4289	1.8014
MEAN	574.42	2.8550	108.63	1.6167	2.0167
UP 95% CI	590.74	3.0214	130.06	1.8045	2.2320
SD	63.206	0.6443	82.946	0.7270	0.8334
VARIANCE	3995.0	0.4151	6880.1	0.5285	0.6946
SE MEAN	8.1599	0.0832	10.708	0.0939	0.1076
C.V.	11.004	22.566	76.354	44.969	41.328

	SAT	GPA	Y	EX	JUNK
MINIMUM	500.00	1.4000	40.000	0.5000	0.0000
1ST QUARTI	520.00	2.5000	70.000	1.0000	2.0000
MEDIAN	565.00	2.9500	80.000	1.5000	2.0000
3RD QUARTI	610.00	3.2000	120.00	2.0000	2.7500
MAXIMUM	780.00	4.0000	500.00	4.0000	4.0000
SKEW	1.0516	-0.0827	3.2006	1.1561	-0.0310
KURTOSIS	0.8106	-0.5764	11.111	0.8912	0.1793

∧∧

(b) Description of inferential differences between female (F) and male (M) students (where S.D. = standard deviation, S.E. = standard error, T = t-statistic, DF = degrees of freedom, P = P-value, CI = confidence interval, F = F statistic, NUM = numerator, DEN = denominator):

(b1) GPA:

```
          TWO-SAMPLE T TESTS FOR GPA BY SEX

                        SAMPLE
   SEX         MEAN      SIZE       S.D.        S.E.
----------  ---------  ------    -------      ------
F            3.0500      60       0.5792      0.0748
M            2.8550      60       0.6443      0.0832
DIFFERENCE   0.1950

NULL HYPOTHESIS: DIFFERENCE =  0
ALTERNATIVE HYP: DIFFERENCE <> 0

ASSUMPTION            T    DF    P    95% CI FOR DIFF.
-----------------    ---  ----  ----  ----------------
EQUAL VARIANCES     1.74  118   0.0838 (-0.0265,0.4165)
UNEQUAL VARIANCES   1.74  116.7 0.0839 (-0.0265,0.4165)

                      F    NUM DF   DEN DF    P
TESTS FOR EQUALITY   -----  ------   ------   -----
      OF VARIANCES   1.24     59       59    0.2078

CASES INCLUDED 120    MISSING CASES 0
```

Based on the above results, the null hypothesis that the mean GPA of female students is equal to the mean GPA of female students cannot be rejected at a P-value ≤ 0.05. Similarly for variances.

∧∧

∧∧

(b2) SAT:

```
            TWO-SAMPLE T TESTS FOR SAT BY SEX

                        SAMPLE
    SEX        MEAN      SIZE       S.D.        S.E.
--------    --------    ------    ---------    ---------
F           595.83        60       54.712       7.0632
M           574.42        60       63.206       8.1599
DIFFERENCE  21.417

NULL HYPOTHESIS: DIFFERENCE =  0
ALTERNATIVE HYP: DIFFERENCE <> 0

ASSUMPTION             T    DF     P    95% CI FOR DIFF.
-----------------     ----  -----  -----  --------------
EQUAL VARIANCES       1.98  118   0.0495  (0.0451,42.788)
UNEQUAL VARIANCES     1.98  115.6 0.0496  (0.0406,42.793)

                       F     NUM DF   DEN DF      P
TESTS  FOR EQUALITY  -----   ------   ------   ------
       OF VARIANCES  1.33      59       59     0.1352

CASES INCLUDED 120     MISSING CASES 0
```

Based on the above results, the null hypothesis that the mean SAT of female students is equal to the mean SAT of male students is rejected because the P-value = 0.0495 (or, 0.0496.) The null hypothesis that the variance of SAT for female students is equal to the variance of SAT for male students cannot be rejected at a P-value ≤ 0.05.

∧∧

(b3) HOURS OF EXERCISE (EX) PER DAY:

```
            TWO-SAMPLE T TESTS FOR EX BY SEX

                        SAMPLE
    SEX        MEAN      SIZE       S.D.        S.E.
--------    --------    ------    ---------    -------
F           1.2583        60       0.5930       0.0766
M           1.6167        60       0.7270       0.0939
DIFFERENCE  -0.3583

NULL HYPOTHESIS: DIFFERENCE =  0
ALTERNATIVE HYP: DIFFERENCE <> 0
```

```
ASSUMPTION              T    DF     P    95% CI FOR DIFF.
----------------      -----  -----  ----  -----------------
EQUAL VARIANCES    -2.96  118     0.0037(-0.5982,-0.1185)
UNEQUAL VARIANCES-2.96  113.4  0.0038(-0.5983,-0.1184)

                        F      NUM DF   DEN DF     P
TESTS FOR EQUALITY    ------  ------   ------   ----
        OF VARIANCES   1.50     59       59    0.0602

CASES INCLUDED 120     MISSING CASES 0
```

Based on the above results, the null hypothesis that the mean EX of female students is equal to the mean EX of male students is rejected because the P-value = 0.0037 (or, 0.0038.) The null hypothesis that the variance of EX for female students is equal to the variance of EX for male students cannot be rejected at a P-value ≤ 0.05.

∧∧∧

(b4) SPENDING ON JUNK FOOD (JUNK):

```
        TWO-SAMPLE T TESTS FOR JUNK BY SEX

                        SAMPLE
   SEX        MEAN       SIZE      S.D.         S.E.
----------  --------    ------   ----------   --------
F           1.4333        60      0.6979       0.0901
M           2.0167        60      0.8334       0.1076
DIFFERENCE  -0.5833

NULL HYPOTHESIS: DIFFERENCE =  0
ALTERNATIVE HYP: DIFFERENCE <> 0

ASSUMPTION              T    DF     P    95% CI FOR DIFF.
----------------      -----  -----  ----  -----------------
EQUAL VARIANCES    -4.16  118     0.0001(-0.8612,-0.3054)
UNEQUAL VARIANCES-4.16  114.5  0.0001(-0.8613,-0.3053)

                        F      NUM DF   DEN DF     P
TESTS FOR EQUALITY    ------  ------   ------   -----
        OF VARIANCES   1.43     59       59    0.0878

CASES INCLUDED 120     MISSING CASES 0
```

Based on the above results, the null hypothesis that the mean JUNK spending of female students is equal to the mean JUNK spending of male students is rejected because the P-value = 0.0001. The null hypothesis that the variance of JUNK for female students is equal to the variance of JUNK for male students cannot be rejected at a P-value ≤ 0.05.

∧∧∧

(b.5) INCOME (INC):

```
              TWO-SAMPLE T TESTS FOR INC BY SEX

                         SAMPLE
     SEX        MEAN       SIZE        S.D.        S.E.
  ----------  ---------  -------    ---------    ------
  F            121.18       60       53.145      6.8609
  M            108.63       60       82.946      10.708
  DIFFERENCE   12.550

  NULL HYPOTHESIS: DIFFERENCE =  0
  ALTERNATIVE HYP: DIFFERENCE <> 0

  ASSUMPTION           T     DF    P      95% CI FOR DIFF.
  -----------------   ----  ----  ----   ----------------
  EQUAL VARIANCES     0.99  118    0.3258 (-12.635,37.735)
  UNEQUAL VARIANCES   0.99 100.5   0.3261 (-12.680,37.780)

                         F     NUM DF   DEN DF     P
  TESTS FOR EQUALITY  -------  ------   ------   ----
         OF VARIANCES  2.44      59       59    0.0004

  CASES INCLUDED 120      MISSING CASES 0
```

Based on the above results, the null hypothesis that the mean INC of female students is equal to the mean INC of male students cannot be rejected at a P-value ≤ 0.05. The null hypothesis that the variance of INC for female students is equal to the variance of INC for male students is rejected because P-value = 0.0004.

∧∧∧

(c) Display of a correlation matrix for variables GPA, INC, SAT, JUNK and EX along with corresponding P-values.

Correlations matrix for variables GPA, INC, SAT, JUNK AND EX:

```
                  CORRELATIONS  (PEARSON)

            INC        GPA        SAT       JUNK
  GPA      0.5691
  P-VALUE  0.0000

  SAT      0.5948     0.8381
           0.0000     0.0000
```

	INC	GPA	SAT	JUNK
JUNK	-0.0293	-0.1223	0.0470	
	0.7507	0.1834	0.6104	
EX	-0.2412	-0.5266	-0.5767	-0.0908
	0.0080	0.0000	0.0000	0.3240

CASES INCLUDED 120   MISSING CASES 0

GPA is related positively and significantly to INC and SAT; negatively and in-significantly to JUNK and negatively and significantly to EX.

SAT is related positively and significantly to INC and GAP; positively and in-significantly to JUNK and negatively and significantly to EX.

∧∧

(d) Display of regressions and scatterplots. Regressions and scatterplots (P-values are reported in parentheses):

(d1) GPA vs INC (It is shown that the results for SAT vs INC are similar.)

$$\text{GPA} = \underset{(0.00)}{2.372} + \underset{(0.00)}{0.005(\text{INC})}; \quad R^2 = 0.32$$

The theory is supported by the findings: the estimated coefficients have the right signs and both are statistically significant. INC explains 32% of GPA's variation, not atypical in cross-section studies.

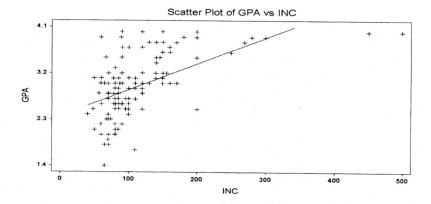

The results for SAT vs INC are similar:

SAT=   526.411   +   0.511(INC) ; $R^2 = 0.35$
        (0.00)   (0.00)

The theory is supported by the findings: the estimated coefficients have the right signs and both are statistically significant. INC explains 35% of GPA's variation, not atypical in cross-section studies.

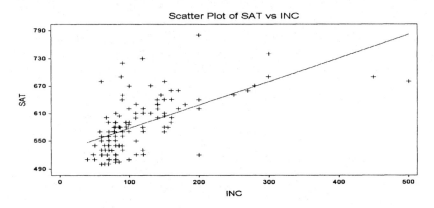

Scatter Plot of SAT vs INC

‿‿

(d2) GPA vs EX

GPA=   3.636   -   0.475(EX) ; $R^2 = 0.28$
        (0.00)   (0.00)

The theory is supported by the findings: the estimated coefficients have the right signs and both are statistically significant. EX explains 28% of GPA's variation, not atypical in cross-section studies.

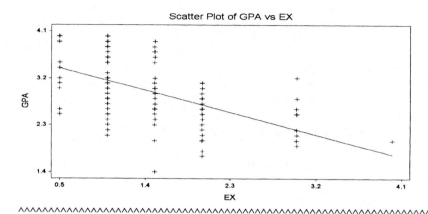

Scatter Plot of GPA vs EX

(d3) GPA vs JUNK

$$GPA = 3.111 \quad - \quad 0.092(JUNK); \qquad R^2 = 0.02$$
$$(0.00) \qquad (0.18)$$

Although the estimated coefficients have the right signs, the P-value of the slope coefficient is insignificant and the R^2 is nearly zero. Hence, the results do not support the theory.

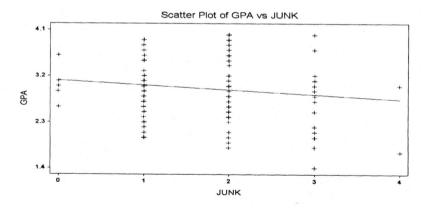

Scatter Plot of GPA vs JUNK

(d4) (GPA) vs (INC, EX, JUNK)

$$GPA = 3.230 + 0.004(INC) - 0.387(EX) - 0.111(JUNK);$$
$$(0.00) \quad (0.00) \qquad (0.00) \qquad (0.03)$$
$$adjR^2 = 0.49, F_{st} = 39.67$$
$$(0.00)$$

The theory is supported by the findings: the estimated coefficients have the right signs and all are statistically significant. The independent variables explain about 49% of GPA's variation which for a cross-section study is very good. The F-statistic is significant, thus the regression makes sense.

∧∧

(e) Display of proportions of political affiliation (PA) by Sex (SEX) along with one and two parameter tests:

(e1) Cross Tabulation of PA by SEX.

PA	SEX F	M	TOTAL
D	29	22	51
I	16	10	26
O	6	7	13
R	9	21	30
TOTAL	60	60	120

∧∧

(e2) Test, at 5%, the claim that the population proportion of democrats is equal to or greater than 55% and a 95% confidence interval about the population proportion.

```
ONE-SAMPLE PROPORTION TEST

SAMPLE SIZE              120
SUCCESSES                51
PROPORTION          0.42500

NULL HYPOTHESIS: P = 0.55
ALTERNATIVE HYP: P < 0.55

DIFFERENCE         -0.12500
STANDARD ERROR      0.04513
Z        -2.75      P   0.0030

95% CONFIDENCE INTERVAL (0.33655, 0.51345)
```

Reject the claim.

∧∧

^^

(e3) Test, at 5%, the claim that the population proportion of non-democrats is greater than 60% and a 95% confidence interval about the population proportion.

```
ONE-SAMPLE PROPORTION TEST

SAMPLE SIZE            120
SUCCESSES               69
PROPORTION          0.57500

NULL HYPOTHESIS: P = 0.6
ALTERNATIVE HYP: P > 0.6

DIFFERENCE          -0.02500
STANDARD ERROR       0.04513
Z        -0.56     P   0.7119

95% CONFIDENCE INTERVAL (0.48655, 0.66345)
```

The H_0 cannot be rejected and so is the claim.

^^

(e4) Test, at 5%, the claim that the population proportion of female democrats is equal to the population proportion of male democrats and a 95% confidence interval about the difference in the population proportions.

```
TWO-SAMPLE PROPORTION TEST

                   SAMPLE 1    SAMPLE 2
SAMPLE SIZE           60          60
SUCCESSES             29          22
PROPORTION         0.48333     0.36667

NULL HYPOTHESIS: P1 =  P2
ALTERNATIVE HYP: P1 <> P2

DIFFERENCE         0.11667
SE (DIFF)          0.09025
Z        1.29     P   0.1961

95% CONFIDENCE INTERVAL OF DIFFERENCE
(-0.06023, 0.29356)
```

The H_0 cannot be rejected. Thus, the claim is rejected.

^^

∧∧∧

(e5) Test, at 5%, the claim that the population proportion of female non-democrats is equal to the population proportion of male non-democrats and a 95% confidence interval about the difference in the population proportions.

TWO-SAMPLE PROPORTION TEST

	SAMPLE 1	SAMPLE 2
SAMPLE SIZE	60	60
SUCCESSES	31	38
PROPORTION	0.51667	0.63333

NULL HYPOTHESIS: P1 = P2
ALTERNATIVE HYP: P1 <> P2

DIFFERENCE   -0.11667
SE (DIFF)   0.09025
Z   -1.29   P 0.1961

95% CONFIDENCE INTERVAL OF DIFFERENCE
(-0.29356, 0.06023)

The H_0 cannot be rejected. Thus, the claim is rejected.

∧∧∧

(f) Sample size.

The sample size, average, standard deviation, standard error (S.E.), 95% confidence interval and margin of error or maximum error of the estimate (E) for SAT and GPA for all students and by sex are:

DESCRIPTIVE STATISTICS FOR ALL STUDENTS

	SAT	GPA
n	120	120
LO 95% CI	574.31	2.8408
MEAN	585.13	2.9525
UP 95% CI	595.94	3.0642
SD	59.837	0.6178
SE	5.4623	0.0564
E	10.706	0.1105

```
DESCRIPTIVE STATISTICS FOR SEX = F

                        SAT            GPA

n                        60             60
LO 95% CI             581.70         2.9004
MEAN                  595.83         3.0500
UP 95% CI             609.97         3.1996
SD                    54.712         0.5792
SE                    7.0632         0.0748
E                     13.844         0.1466

DESCRIPTIVE STATISTICS FOR SEX = M

                        SAT            GPA

n                        60             60
LO 95% CI             558.09         2.6886
MEAN                  574.42         2.8550
UP 95% CI             590.74         3.0214
SD                    63.206         0.6443
SE                    8.1599         0.0832
E                     15.993         0.1631
```

^^^

(f1) If we wish to construct a 95% confidence interval about the mean of the SAT college population, what should the sample size be with E = 5?

$$n = [(59.837)(1.96)/5]^2 \approx 550$$

^^^

(f2) If we wish to construct a 95% confidence interval about the mean of the GPA college population, what should the sample size be with E = .08?

$$n = [(0.6178)(1.96)/0.08]^2 \approx 229$$

^^^

(f3) If we wish to construct a 95% confidence interval about the mean population difference ($SAT_{Female} - SAT_{Male}$), what should the sample size be with E = 5?

$$n = [(1.96)^2(54.712^2 + 63.206^2)] / 5^2 \approx 1074$$

^^^

^^

(f4) If we wish to construct a 95% confidence interval about the mean population difference $(GPA_{Female} - GPA_{Male})$, what should the sample size be with $E = .08$?

$$n = [(1.96)^2(0.5792^2 + 0.6443^2)] / 0.08^2 \approx 451$$

^^

(f5) If we wish to construct a 95% confidence interval about the population proportion of college democrats, what should the sample size be with $E = .05$?

$$n = (0.5)(0.5)(1.96/.05)^2 \approx 384$$

^^

(f6) If we wish to construct a 99% confidence interval about the population proportion of female college democrats, what should the sample size be with $E = .05$?

$$n = (0.5)(0.5)(2.575/0.05)^2 \approx 663$$

^^

(f7) If we wish to construct a 95% confidence interval about the proportion population difference $(Democrat_{Female} - Democrat_{Male})$, what should the sample size be with $E = .05$?

$$n = (1.96)^2[(0.48)(0.52)+(0.37)(0.63)] / 0.05^2 \approx 742$$

^^

(f8) Test at 95% of the claim that the men SAT population average is equal to or greater than the female SAT population average.

Test:
$$H_0: \mu \geq 595.83$$
$$H_1: \mu < 595.83$$

The H_0 is rejected (and so is the claim) because

$$Z_\alpha = -1.645 > Z_{st} = (574.42-595.830)/(8.16) = -2.62.$$

Power of Test:

Let $\mu_\tau = 580$. Therefore,

$$\beta = P[Z \geq (595.83-580)/(8.16) - |Z_\alpha|] = P[Z \geq 0.295] = 0.3859$$

$$\text{Power} = 1 - \beta = 0.6141.$$

Thus, there is a 61.41% chance of rejecting the H_0 given that the H_0 is false and the true population mean is 580.

Size of Sample for $\alpha = 0.05$ and Power = 0.95 or $\beta = 0.05$:

$$n = [(Z_\alpha + Z_\beta)^2 s^2] / \Delta^2 = [(1.645+1.645)^2 (63.206^2)] / 15.83^2 \approx 173$$

Size of Sample for $\alpha = 0.05$ and Power = 0.99 or $\beta = 0.01$:

$$n = [(Z_\alpha + Z_\beta)^2 s^2] / \Delta^2 = [(1.645+2.33)^2 (63.206^2)] / 15.83^2 \approx 252.$$

Thus, for a 99% chance of rejecting the H_0 given that the H_0 is false and the true population mean is 580, the sample size should be $n \approx 252$.

∧∧

Appendix II:

SUMMATION NOTATION

Consider the set of numbers: $y_1, y_2, \ldots y_n$. Their sum may be represented by:

$$\sum_{i=1}^{n} y_i = y_1 + y_2 + \ldots y_n.$$

Note: $\sum$ (read "capital sigma") means "the sum of," i (or any letter) is the index of summation, and numbers 1 and n are the lower and upper limits of the summation respectively.

e.g. Consider a set y which consists of numbers: $y_1 = 2$, $y_2 = 3$, $y_3 = 4$, and $y_4 = 5$. Then,

$$\sum_{i=1}^{4} y_i = 2 + 3 + 4 + 5 = 14$$

If we would like to consider some of y's subsets, then,

$$\sum_{i=2}^{4} y_i = 3 + 4 + 5 = 12 \quad \text{or} \quad \sum_{i=1}^{3} y_i = 2 + 3 + 4 = 9$$

If k is a constant, then,

$$\sum_{i=1}^{n} ky_i = \sum_{i=1}^{n} y_i$$

e.g. Consider set y, introduced above, the elements of which are now multiplied by 2. Then,

$$\sum_{i=1}^{4} 2y_i = 2 \sum_{i=1}^{4} y_i = 2(2 + 3 + 4 + 5) = 28$$

Similarly,

$$\sum_{i=1}^{n} k = nk, \sum_{i=1}^{n} (y_i + k) = \sum_{i=1}^{n} y_i + nk \text{ and}$$

$$\sum_{i=1}^{n} (cy_i + dx_i) = c \sum_{i=1}^{n} y_i + d \sum_{i=1}^{n} x_i .$$

e.g. Consider set z which consists of 4, 4, 4, and 4. Consider set y, introduced above, and let set x, consist of 20, 30, 40, and 50. Then,

$$\sum_{i=1}^{4} 4 = (4)(4) = 16 , \sum_{i=1}^{4} (y + 4) = 14 + 16 = 30 ,$$

$$\sum_{i=1}^{4} (cy_i + dx_i) = 14c + 140d$$

Appendix III: Tables

TABLE III.1a

STANDARIZED NORMAL DISTRIBUTION

z	0.00	0.01	0.02	0.03	0.04
0.0	0.5000	0.4960	0.4920	0.4880	0.4840
0.1	0.4602	0.4562	0.4522	0.4483	0.4443
0.2	0.4207	0.4168	0.4129	0.4090	0.4052
0.3	0.3821	0.3873	0.3745	0.3707	0.3669
0.4	0.3446	0.3409	0.3372	0.3336	0.3300
0.5	0.3085	0.3050	0.3015	0.2981	0.2946
0.6	0.2743	0.2709	0.2676	0.2643	0.2611
0.7	0.2420	0.2389	0.2358	0.2327	0.2296
0.8	0.2119	0.2090	0.2061	0.2033	0.2005
0.9	0.1841	0.1814	0.1788	0.1762	0.1736
1.0	0.1587	0.1562	0.1539	0.1515	0.1492
1.1	0.1357	0.1335	0.1314	0.1292	0.1271
1.2	0.1151	0.1131	0.1112	0.1093	0.1075
1.3	0.0968	0.0951	0.0934	0.0918	0.0901
1.4	0.0808	0.0793	0.0778	0.0764	0.0749
1.5	0.0668	0.0655	0.0643	0.0630	0.0618
1.6	0.0548	0.0537	0.0526	0.0516	0.0505
1.7	0.0446	0.0436	0.0427	0.0418	0.0409
1.8	0.0359	0.0351	0.0344	0.0366	0.0329
1.9	0.0287	0.0281	0.0274	0.0268	0.0262
2.0	0.0228	0.0222	0.0217	0.0212	0.0207
2.1	0.0179	0.0174	0.0170	0.0166	0.0162
2.2	0.0139	0.0136	0.0132	0.0129	0.0125
2.3	0.0107	0.0104	0.0102	0.0099	0.0096
2.4	0.0082	0.0080	0.0078	0.0075	0.0073
2.5	0.0062	0.0060	0.0059	0.0057	0.0055
2.6	0.0047	0.0045	0.0044	0.0043	0.0041
2.7	0.0035	0.0034	0.0033	0.0032	0.0031
2.8	0.0026	0.0025	0.0024	0.0023	0.0023

TABLE III.1b

STANDARIZED NORMAL DISTRIBUTION

Z	0.05	0.06	0.07	0.08	0.09
0.0	0.4801	0.4761	0.4721	0.4681	0.4641
0.1	0.4404	0.4364	0.4325	0.4686	0.4247
0.2	0.4013	0.3974	0.3936	0.3897	0.3859
0.3	0.3632	0.3594	0.3557	0.3520	0.3483
0.4	0.3264	0.3228	0.3192	0.3156	0.3121
0.5	0.2912	0.2877	0.2843	0.2810	0.2776
0.6	0.2578	0.2546	0.2514	0.2483	0.2451
0.7	0.2266	0.2236	0.2206	0.2217	0.2148
0.8	0.1977	0.1949	0.1922	0.1894	0.1867
0.9	0.1711	0.1685	0.1660	0.1635	0.1611
1.0	0.1469	0.1446	0.1423	0.1401	0.1379
1.1	0.1251	0.1230	0.1210	0.1190	0.1170
1.2	0.1056	0.1038	0.1020	0.1003	0.0985
1.3	0.0885	0.0869	0.0853	0.0838	0.0823
1.4	0.0735	0.0721	0.0708	0.0694	0.0681
1.5	0.0606	0.0594	0.0582	0.0571	0.0559
1.6	0.0495	0.0485	0.0475	0.0465	0.0455
1.7	0.0401	0.0392	0.0384	0.0375	0.0367
1.8	0.0322	0.0314	0.0307	0.0301	0.0294
1.9	0.0256	0.0250	0.0244	0.0239	0.0233
2.0	0.0202	0.0197	0.0192	0.0188	0.0183
2.1	0.0158	0.0154	0.0150	0.0146	0.0143
2.2	0.0122	0.0119	0.0116	0.0113	0.0110
2.3	0.0094	0.0091	0.0089	0.0087	0.0084
2.4	0.0071	0.0069	0.0068	0.0066	0.0064
2.5	0.0054	0.0052	0.0051	0.0049	0.0048
2.6	0.0040	0.0039	0.0038	0.0037	0.0036
2.7	0.0030	0.0029	0.0028	0.0027	0.0026
2.8	0.0022	0.0021	0.0020	0.0020	0.0019

TABLE III.2

PERCENTILES OF THE t DISTRIBUTION
Level of significance for one-tailed test

df	0.10	0.05	0.025	0.01	0.005	0.0005
		Level of significance for two-tailed test				
	0.20	0.10	0.05	0.02	0.01	0.001
1	3.078	6.314	12.706	31.821	63.657	636.619
2	1.886	2.920	4.303	6.965	9.925	31.599
3	1.638	2.353	3.182	4.541	5.841	12.924
4	1.533	1.132	2.776	3.747	4.604	8.610
5	1.476	2.015	2.571	3.365	4.032	6.869
6	1.440	1.943	2.447	3.143	3.707	5.959
7	1.415	1.895	2.365	2.998	3.499	5.408
8	1.397	1.860	2.306	2.896	3.355	5.041
9	1.383	1.833	2.262	2.821	3.250	4.781
10	1.372	1.812	2.228	2.764	3.169	4.587
11	1.363	1.796	2.201	2.718	3.106	4.437
12	1.356	1.782	2.179	2.681	3.055	4.318
13	1.350	1.771	2.160	2.650	3.012	4.221
14	1.345	1.761	2.145	2.624	2.977	4.140
15	1.341	1.753	2.131	2.602	2.947	4.073
16	1.337	1.746	2.120	2.583	2.921	4.015
17	1.333	1.740	2.110	2.567	2.898	3.965
18	1.330	1.734	2.101	2.552	2.878	3.922
19	1.328	1.729	2.093	2.539	2.861	3.883
20	1.325	1.725	2.086	2.528	2.845	3.850
21	1.323	1.721	2.080	2.518	2.831	3.819
22	1.321	1.717	2.074	2.508	2.819	3.792
23	1.319	1.714	2.069	2.500	2.807	3.768
24	1.318	1.711	2.064	2.492	2.797	3.745
25	1.316	1.708	2.060	2.485	2.787	3.725
26	1.315	1.706	2.056	2.479	2.779	3.707
27	1.314	1.703	2.052	2.473	2.771	3.690
28	1.313	1.701	2.048	2.467	2.763	3.674
29	1.311	1.699	2.045	2.462	2.756	3.659
30	1.310	1.697	2.042	2.457	2.750	3.646
40	1.303	1.684	2.021	2.423	2.704	3.551
60	1.296	1.671	2.000	2.390	2.660	3.460
120	1.289	1.658	1.980	2.358	2.617	3.373
ж	1.282	1.645	1.960	2.326	2.576	3.291

TABLE III.3a

PERCENTILES OF THE X^2 DISTRIBUTION

df	\multicolumn{5}{c}{Percent}
	0.5	1	2.5	5	10
1	0.000039	0.00016	0.00098	0.0039	0.0158
2	0.0100	0.0201	0.0506	0.1026	0.2107
3	0.0717	0.115	0.216	0.352	0.584
4	0.207	0.297	0.484	0.711	1.064
5	0.412	0.554	0.831	1.15	1.61
6	0.676	0.872	1.24	1.64	2.20
7	0.989	1.24	1.69	2.17	2.83
8	1.34	1.65	2.18	2.73	3.49
9	1.73	2.09	2.70	3.33	4.17
10	2.16	2.56	3.25	3.94	4.87
11	2.60	3.05	3.82	4.57	5.58
12	3.07	3.57	4.40	5.23	6.30
13	3.57	4.11	5.01	5.89	7.04
14	4.07	4.66	5.63	6.57	7.79
15	4.60	5.23	6.26	7.26	8.55
16	5.14	5.81	6.91	7.96	9.31
18	6.26	7.01	8.23	9.39	10.86
20	7.43	8.26	9.59	10.85	12.44
24	9.89	10.86	12.40	13.85	15.66
30	13.79	14.95	16.79	18.49	20.60
40	20.71	22.16	24.43	26.51	29.05
60	35.53	37.48	40.48	43.19	46.46
120	83.85	86.92	91.58	95.70	100.62

TABLE III.3b

PERCENTILES OF THE X^2 DISTRIBUTION

df	Percent				
	90	95	97.5	99	99.5
1	2.71	3.84	5.02	6.63	7.88
2	4.61	5.99	7.38	9.21	10.60
3	6.25	7.81	9.35	11.34	12.84
4	7.78	9.49	11.14	13.28	14.86
5	9.24	11.07	12.83	15.09	16.75
6	10.64	12.59	14.45	16.81	18.55
7	12.02	14.07	16.01	18.48	20.28
8	13.36	15.51	17.53	20.09	21.96
9	14.68	16.92	19.02	21.67	23.59
10	15.99	18.31	20.48	23.21	25.19
11	17.28	19.68	21.92	24.73	26.76
12	18.55	21.03	23.34	26.22	28.30
13	19.81	22.36	24.74	27.69	29.82
14	21.06	23.68	26.12	29.14	31.32
15	22.31	25.00	27.49	30.58	32.80
16	23.54	26.30	28.85	32.00	34.27
18	25.99	28.87	31.53	34.81	37.16
20	28.41	31.41	34.17	37.57	40.00
24	33.20	36.42	39.36	42.98	45.56
30	40.26	43.77	47.98	50.89	53.67
40	51.81	55.76	59.34	26.51	66.77
60	74.40	79.08	83.30	43.19	91.95
120	140.23	146.57	152.21	95.70	163.64

Appendix III

TABLE III.4a

F DISTRIBUTION, 5 PERCENT SIGNIFICANCE
Degrees of freedom for numerator

df	1	2	3	4	5	6	7	8	9	10
1	161	200	216	225	230	234	237	239	241	242
2	18.5	19.0	19.2	19.2	19.3	19.3	19.4	19.4	19.4	19.4
3	10.1	9.55	9.28	9.12	9.01	8.94	8.89	8.85	8.81	8.79
4	7.71	6.94	6.59	6.39	6.26	6.16	6.09	6.04	6.00	5.96
5	6.61	5.79	5.41	5.19	5.05	4.95	4.88	4.82	4.77	4.74
6	5.99	5.14	4.76	4.53	4.39	4.28	4.21	4.15	4.10	4.06
7	5.59	4.74	4.35	4.12	3.97	3.87	3.79	3.73	3.68	3.64
8	5.32	4.46	4.07	3.84	3.69	3.58	3.50	3.44	3.39	3.35
9	5.12	4.26	3.86	3.63	3.48	3.37	3.29	3.23	3.18	3.14
10	4.96	4.10	3.71	3.48	3.33	3.22	3.14	3.07	3.02	2.98
11	4.84	3.98	3.59	3.36	3.20	3.09	3.01	2.95	2.90	2.85
12	4.75	3.89	3.49	3.26	3.11	3.00	2.91	2.85	2.80	2.75
13	4.67	3.81	3.41	3.18	3.03	2.92	2.83	2.77	2.71	2.67
14	4.60	3.74	3.34	3.11	2.96	2.85	2.76	2.70	2.65	2.60
15	4.54	3.68	3.29	3.06	2.90	2.79	2.71	2.64	2.59	2.54
16	4.49	3.63	3.24	3.01	2.85	2.74	2.66	2.59	2.54	2.49
17	4.45	3.59	3.20	2.96	2.81	2.70	2.61	2.55	2.48	2.45
18	4.41	3.55	3.16	2.93	2.77	2.66	2.58	2.51	2.46	2.41
19	4.38	3.52	3.13	2.90	2.74	2.63	2.54	2.48	2.42	2.39
20	4.35	3.49	3.10	2.87	2.71	2.60	2.51	2.45	2.39	2.35
21	4.32	3.47	3.07	2.84	2.68	2.57	2.49	2.42	2.37	2.32
22	4.30	3.44	3.05	2.82	2.66	2.55	2.46	2.40	2.34	2.30
23	4.28	3.42	3.03	2.80	2.64	2.53	2.44	2.37	2.32	2.27
24	4.26	3.40	3.01	2.78	2.62	2.51	2.42	2.36	2.30	2.25
25	4.24	3.39	2.99	2.76	2.60	2.49	2.40	2.34	2.28	2.24
30	4.17	3.32	2.92	2.69	2.53	2.42	2.33	2.27	2.21	2.16
40	4.08	3.23	2.84	2.61	2.45	2.34	2.25	2.18	2.12	2.08
60	4.00	3.15	2.76	2.53	2.37	2.25	2.17	2.10	2.04	1.99
120	3.92	3.07	2.68	2.45	2.29	2.18	2.09	2.02	1.96	1.91
∞	3.84	3.00	2.60	2.37	2.21	2.10	2.01	1.94	1.88	1.83

TABLE III.4b

F DISTRIBUTION, 5 PERCENT SIGNIFICANCE
Degrees of freedom for numerator

df	12	15	20	24	30	40	60	120	∞
1	244	246	248	249	250	251	252	253	254
2	19.4	19.4	19.5	19.5	19.5	19.5	19.5	19.5	19.5
3	8.74	8.70	8.66	8.64	8.62	8.59	8.57	8.55	8.53
4	5.91	5.86	5.80	5.77	5.75	5.72	5.69	5.66	5.63
5	4.68	4.62	4.56	4.53	4.50	4.46	4.43	4.40	4.37
6	4.00	3.94	3.87	3.84	3.81	3.77	3.74	3.70	3.67
7	3.57	3.51	3.44	3.41	3.38	3.34	3.30	3.27	3.23
8	3.28	3.22	3.15	3.12	3.08	3.04	3.01	2.97	2.93
9	3.07	3.01	2.94	2.90	2.86	2.83	2.79	2.75	2.71
10	2.91	2.85	2.77	2.74	2.70	2.66	2.62	2.58	2.54
11	2.79	2.72	2.65	2.61	2.57	2.53	2.49	2.45	2.40
12	2.69	2.62	2.54	2.51	2.47	2.43	2.38	2.34	2.30
13	2.60	2.53	2.46	2.42	2.38	2.34	2.30	2.25	2.21
14	2.53	2.46	2.39	2.35	2.31	2.27	2.22	2.18	2.13
15	2.48	2.40	2.33	2.29	2.25	2.20	2.16	2.11	2.07
16	2.42	2.35	2.28	2.24	2.19	2.15	2.11	2.06	2.01
17	2.38	2.31	2.23	2.19	2.15	2.10	2.06	2.01	1.96
18	2.34	2.27	2.19	2.15	2.11	2.06	2.02	1.97	1.92
19	2.31	2.23	2.16	2.11	2.07	2.03	1.98	1.93	1.88
20	2.28	2.20	2.12	2.08	2.04	1.99	1.95	1.90	1.84
21	2.25	2.18	2.10	2.05	2.01	1.96	1.92	1.87	1.81
22	2.23	2.15	2.07	2.03	1.98	1.94	1.89	1.84	1.78
23	2.20	2.13	2.05	2.01	1.96	1.91	1.86	1.81	1.76
24	2.18	2.11	2.03	1.98	1.94	1.89	1.84	1.79	1.73
25	2.16	2.09	2.01	1.96	1.92	1.87	1.82	1.77	1.71
30	2.09	2.01	1.93	1.89	1.84	1.79	1.74	1.68	1.62
40	2.00	1.92	1.84	1.79	1.74	1.69	1.64	1.58	1.51
60	1.92	1.84	1.75	1.70	1.65	1.59	1.53	1.47	1.39
120	1.83	1.75	1.66	1.61	1.55	1.50	1.43	1.35	1.25
∞	1.75	1.67	1.57	1.52	1.46	1.39	1.32	1.22	1.00

Appendix III

TABLE III.4c

F DISTRIBUTION, 1 PERCENT SIGNIFICANCE

Degrees of freedom for numerator

df	1	2	3	4	5	6	7	8	9	10
1	4052	5000	5403	5625	5746	5859	5928	5982	6023	6056
2	98.5	99.0	99.2	99.2	99.3	99.3	99.4	99.4	99.4	99.4
3	34.1	30.8	29.5	28.7	28.2	27.9	27.7	27.5	27.3	27.2
4	21.2	18.0	16.7	16.0	15.5	15.2	15.0	14.8	14.7	14.5
5	16.3	13.3	12.1	11.4	11.0	10.7	10.5	10.3	10.2	10.1
6	13.7	10.9	9.78	9.15	8.75	8.47	8.26	8.10	7.98	7.87
7	12.2	9.55	8.45	7.85	7.46	7.19	6.99	6.84	6.72	6.62
8	11.3	8.65	7.59	7.01	6.63	6.37	6.18	6.03	5.91	5.81
9	10.6	8.02	6.99	6.42	6.06	5.80	5.61	5.47	5.35	5.26
10	10.0	7.56	6.55	5.99	5.64	5.39	5.20	5.06	4.94	4.85
11	9.65	7.21	6.22	5.67	5.32	5.07	4.89	4.74	4.63	4.54
12	9.33	6.93	5.95	5.41	5.06	4.82	4.64	4.50	4.39	4.30
13	9.07	6.70	5.74	5.21	4.97	4.62	4.44	4.30	4.19	4.10
14	8.86	6.51	5.56	5.04	4.70	4.46	4.28	4.14	4.03	3.94
15	8.68	6.36	5.42	4.89	4.56	4.32	4.14	4.00	3.89	3.80
16	8.53	6.23	5.29	4.77	4.44	4.20	4.03	3.89	3.78	3.69
17	8.40	6.11	5.19	4.67	4.34	4.10	3.93	3.79	3.68	3.59
18	8.29	6.01	5.09	4.58	4.25	4.01	3.84	3.71	3.60	3.51
19	8.19	5.93	5.01	4.50	4.17	3.94	3.77	3.63	3.52	3.43
20	8.10	5.85	4.94	4.43	4.10	3.87	3.70	3.56	3.46	3.37
21	8.02	5.78	4.87	4.37	4.04	3.81	3.64	3.51	3.40	3.31
22	7.95	5.72	4.82	4.31	3.99	3.76	3.59	3.45	3.35	3.26
23	7.88	5.66	4.76	4.26	3.94	3.71	3.54	3.41	3.30	3.21
24	7.82	5.61	4.72	4.22	3.90	3.67	3.50	3.36	3.26	3.17
25	7.77	5.57	4.68	4.18	3.86	3.63	3.46	3.32	3.22	3.13
30	7.56	5.39	4.51	4.02	3.70	3.47	3.30	3.17	3.07	2.98
40	7.31	5.18	4.31	3.83	3.51	3.29	3.12	2.99	2.89	2.80
60	7.08	4.98	4.13	3.65	3.34	3.12	2.95	2.82	2.72	2.63
120	6.85	4.79	3.95	3.48	3.17	2.96	2.79	2.66	2.56	2.47
∞	6.63	4.61	3.78	3.32	3.02	2.80	2.64	2.51	2.41	2.32

TABLE III.4d

F DISTRIBUTION, 1 PERCENT SIGNIFICANCE
Degrees of freedom for numerator

df	12	15	20	24	30	40	60	120	∞
1	6106	6157	6209	6235	6261	6287	6313	6339	6366
2	99.4	99.4	99.4	99.5	99.5	99.5	99.5	99.5	99.5
3	27.1	26.9	26.7	26.6	26.5	26.4	26.3	26.2	26.1
4	14.4	14.2	14.0	13.9	13.8	13.7	13.7	13.6	13.5
5	9.89	9.72	9.55	9.47	9.38	9.29	9.20	9.11	9.02
6	7.72	7.56	7.40	7.31	7.23	7.14	7.06	6.97	6.88
7	6.47	6.31	6.16	6.07	5.99	5.91	5.82	5.74	5.65
8	5.67	5.52	5.36	5.28	5.29	5.12	5.03	4.95	4.86
9	5.11	4.96	4.81	4.73	4.65	4.57	4.48	4.40	4.31
10	4.71	4.56	4.41	4.33	4.25	4.17	4.08	4.00	3.91
11	4.40	4.25	4.10	4.02	3.94	3.86	3.78	3.69	3.60
12	4.18	4.01	3.86	3.78	3.70	3.62	3.54	3.45	3.36
13	3.96	3.82	3.66	3.59	3.51	3.43	3.34	3.25	3.17
14	3.80	3.66	3.51	4.43	3.35	3.27	3.18	3.09	3.00
15	3.67	3.52	3.37	3.29	3.21	3.13	3.05	2.96	2.87
16	3.55	3.41	3.26	3.18	3.10	3.02	2.93	2.84	2.75
17	3.46	3.31	3.16	3.08	3.00	2.92	2.83	2.75	2.65
18	3.37	3.23	3.08	3.00	2.92	2.84	2.75	2.66	2.57
19	3.30	3.15	3.00	2.92	2.84	2.76	2.67	2.58	2.49
20	3.23	3.09	2.94	2.86	2.78	2.68	2.61	2.52	2.42
21	3.17	3.03	2.88	2.80	2.72	2.64	2.55	2.46	2.36
22	3.12	2.98	2.83	2.75	2.67	2.58	2.50	2.40	2.31
23	3.07	2.93	2.78	2.70	2.62	2.54	2.45	2.35	2.26
24	3.03	2.89	2.74	2.66	2.58	2.49	2.40	2.31	2.21
25	2.99	2.85	2.70	2.62	2.53	2.45	2.36	2.27	2.17
30	2.84	2.70	2.55	2.47	2.30	2.39	2.21	2.11	2.01
40	2.66	2.52	2.37	2.29	2.20	2.11	2.02	1.92	1.80
60	2.50	2.35	2.20	2.12	2.03	1.94	1.84	1.73	1.60
120	2.34	2.19	2.03	1.94	1.86	1.76	1.66	1.53	1.38
∞	2.18	2.04	1.88	1.79	1.70	1.59	1.47	1.32	1.00

TABLE III.5

FIVE PERCENT SIGNIFICANCE POINTS OF d_l AND d_u
FOR DURBIN-WATSON TEST**

N	k = 1		k = 2		k = 3		k = 4		k = 5	
	d_l	d_u	d_l	d_u	d_l	d_u	d_l	d_u	d_l	d_u
15	1.08	1.36	0.95	1.54	0.82	1.75	0.69	1.97	0.56	2.21
16	1.10	1.37	0.98	1.54	0.86	1.73	0.74	1.93	0.62	2.15
17	1.13	1.38	1.02	1.54	0.90	1.71	0.78	1.90	0.67	2.10
18	1.16	1.39	1.05	1.53	0.93	1.69	0.82	1.87	0.71	2.06
19	1.18	1.40	1.08	1.53	0.97	1.68	0.86	1.85	0.75	2.02
20	1.20	1.41	1.10	1.54	1.00	1.68	0.90	1.83	0.79	1.99
21	1.22	1.42	1.13	1.54	1.03	1.67	0.93	1.81	0.83	1.96
22	1.24	1.43	1.15	1.54	1.05	1.66	0.96	1.80	0.86	1.94
23	1.26	1.44	1.17	1.54	1.08	1.66	0.99	1.79	0.90	1.92
24	1.27	1.45	1.19	1.55	1.10	1.66	1.01	1.78	0.93	1.90
25	1.29	1.45	1.21	1.55	1.12	1.66	1.04	1.77	0.95	1.89
26	1.30	1.46	1.22	1.55	1.14	1.65	1.06	1.76	0.98	1.88
27	1.32	1.47	1.24	1.56	1.16	1.65	1.08	1.76	1.01	1.86
28	1.33	1.48	1.26	1.56	1.18	1.65	1.10	1.75	1.03	1.85
29	1.34	1.48	1.27	1.56	1.20	1.65	1.12	1.74	1.05	1.84
30	1.35	1.49	1.28	1.57	1.21	1.65	1.14	1.74	1.07	1.83
31	1.36	1.50	1.30	1.57	1.23	1.65	1.16	1.74	1.09	1.83
32	1.37	1.50	1.31	1.57	1.24	1.65	1.18	1.73	1.11	1.82
33	1.38	1.51	1.32	1.58	1.26	1.65	1.19	1.73	1.13	1.81
34	1.39	1.51	1.33	1.58	1.27	1.65	1.21	1.73	1.15	1.81
35	1.40	1.52	1.34	1.53	1.28	1.65	1.22	1.73	1.16	1.80
36	1.41	1.52	1.35	1.59	1.29	1.65	1.24	1.73	1.18	1.80
37	1.42	1.53	1.36	1.59	1.31	1.66	1.25	1.72	1.19	1.80
38	1.43	1.54	1.37	1.59	1.32	1.66	1.26	1.72	1.21	1.79
39	1.43	1.54	1.38	1.60	1.33	1.66	1.27	1.72	1.22	1.79
40	1.44	1.54	1.39	1.60	1.34	1.66	1.29	1.72	1.23	1.79
45	1.48	1.57	1.43	1.62	1.38	1.67	1.34	1.72	1.29	1.78
50	1.50	1.59	1.46	1.63	1.42	1.67	1.38	1.72	1.34	1.77
55	1.53	1.60	1.49	1.64	1.45	1.68	1.41	1.72	1.38	1.77
60	1.55	1.62	1.51	1.65	1.48	1.69	1.44	1.73	1.41	1.77
65	1.57	1.63	1.54	1.66	1.50	1.70	1.47	1.73	1.44	1.77
70	1.58	1.64	1.55	1.67	1.52	1.70	1.49	1.74	1.46	1.77
75	1.60	1.65	1.57	1.68	1.54	1.71	1.51	1.74	1.49	1.77
80	1.61	1.66	1.59	1.69	1.56	1.72	1.53	1.74	1.51	1.77
85	1.62	1.67	1.60	1.70	1.57	1.72	1.55	1.75	1.52	1.77
90	1.63	1.68	1.61	1.70	1.59	1.73	1.57	1.75	1.54	1.78
95	1.64	1.69	1.62	1.71	1.60	1.73	1.58	1.75	1.56	1.78
100	1.65	1.69	1.63	1.72	1.61	1.74	1.59	1.76	1.57	1.78

**N = # of observations; k = # of explanatory variables (excluding the constant term).

Appendix IV:
QUICK-REFERENCE GUIDES

Quick-Reference Guide for Chapter 2
Probability

Population Measure:	Ungrouped (un) Data	Grouped (gr) Data	Raw (r) Data
Mean (μ)	$\Sigma\,[\,X \cdot P(X)\,]$	$\dfrac{\Sigma f \cdot m}{N}$	$\dfrac{\Sigma X}{N}$
Variance (σ^2)	$\Sigma\,(X - \mu_{un})^2 \cdot P(X)$	$\dfrac{\Sigma f \cdot (m - \mu_{gr})^2}{N}$	$\dfrac{\Sigma\,(X - \mu_r)^2}{N}$
Standard Deviation	$\sqrt{\sigma_{un}^2}$	$\sqrt{\sigma_{gr}^2}$	$\sqrt{\sigma_r^2}$
Coefficient of Variation (CV)	$\left(\dfrac{\sigma_{un}}{\mu_{un}}\right)$	$\left(\dfrac{\sigma_{gr}}{\mu_{gr}}\right)$	$\left(\dfrac{\sigma_r}{\mu_r}\right)$
Skewness (π)	$\dfrac{(\mu_{un} - \lambda)}{\sigma_{un}}$	$\dfrac{(\mu_{gr} - \lambda)}{\sigma_{gr}}$	$\dfrac{(\mu_r - \lambda)}{\sigma_r}$
Kurtosis (θ)	$\dfrac{\left[\dfrac{\Sigma(X - \mu_{un})^4}{N}\right]}{\sigma_{un}^4}$	$\dfrac{\left[\dfrac{\Sigma f \cdot (m - \mu_{gr})^4}{N}\right]}{\sigma_{gr}^4}$	$\dfrac{\left[\dfrac{\Sigma(X - \mu_r)^4}{N}\right]}{\sigma_r^4}$

Quick-Reference Guide for Chapter 3
Confidence Intervals

Population Parameter(s):	Confidence Interval	Sample Size
Mean (μ) (when $N \geq 30$)	$P\left[\overline{X} - Z_{\alpha/2}\left(\dfrac{s}{\sqrt{n}}\right) \leq \mu \leq \overline{X} + Z_{\alpha/2}\left(\dfrac{s}{\sqrt{n}}\right)\right]$	$\left[\dfrac{(\sigma)(Z_{\alpha/2})}{E}\right]^2$
Mean (μ) (when $N < 30$)	$P\left[\overline{X} - t_{\alpha/2}\left(\dfrac{s}{\sqrt{n}}\right) \leq \mu \leq \overline{X} + t_{\alpha/2}\left(\dfrac{s}{\sqrt{n}}\right)\right]$	$\left[\dfrac{(\sigma)(Z_{\alpha/2})}{E}\right]^2$
Proportion (ϕ) (when $np \geq 5$ and $n(1-p) \geq 5$)	$P\left[(p - (Z_{\alpha/2})\sigma_p) \leq \phi \leq (p + (Z_{\alpha/2})\sigma_p)\right]$	$Z_{\alpha/2}\left[\sqrt{\dfrac{\phi q}{n}}\right]$
Variance (σ^2)	$P\left[\dfrac{\left[(n-1)s^2\right]}{\chi_L^2} \leq \sigma^2 \leq \dfrac{\left[(n-1)s^2\right]}{\chi_U^2}\right]$	-
Two Means ($\mu_1 - \mu_2$) (when $30 \leq n_1 < .05N_1$ and $30 \leq n_2 < .05N_2$)	$(\overline{X}_1 - \overline{X}_2) \pm Z_{\alpha/2}\sqrt{\dfrac{\sigma_1^2}{n_1} + \dfrac{\sigma_2^2}{n_2}}$	$\dfrac{(Z_{\alpha/2})^2(\sigma_1^2 + \sigma_2^2)}{E^2}$
Two Means ($\mu_1 - \mu_2$) (when $n_1 < 30$ or $n_2 < 30$ and $df = n_1+n_2-2$)	$(\overline{X}_1 - \overline{X}_2) \pm t_{\alpha/2}\sqrt{s_p^2\left(\dfrac{1}{n_1} + \dfrac{1}{n_2}\right)}$	$\dfrac{(Z_{\alpha/2})^2(\sigma_1^2 + \sigma_2^2)}{E^2}$
Two Means ($\mu_1 - \mu_2$) (when $n_1 < 30$ or $n_2 < 30$)	$(\overline{X}_1 - \overline{X}_2) \pm t_{\alpha/2}\sqrt{\dfrac{s_1^2}{n_1} + \dfrac{s_2^2}{n_2}}$	$\dfrac{(Z_{\alpha/2})^2(\sigma_1^2 + \sigma_2^2)}{E^2}$

Two Proportions $(\phi_1 - \phi_2)$ (when $30 \le n_1 < .05N_1$ and $30 \le n_2 < .05N_2$)	$(p_1 - p_2) \pm Z_{\alpha/2} \sqrt{\dfrac{p_1(1-p_1)}{n_1} + \dfrac{p_2(1-p_2)}{n_2}}$	$\dfrac{\left(Z_{\alpha/2}\right)^2 \left(\phi_1 q_1 + \phi_2 q_2\right)}{E^2}$
Health and Medical	$P\left[p - (Z_{\alpha/2})(s) \le Sen_{Pop} \le p + (Z_{\alpha/2})(s) \right]$	$\dfrac{\left(Z_{\alpha/2}^2\right)\left(Sen\right)\left(1 - Sen\right)}{D^2}$

Quick-Reference Guide for Chapter 4: Hypothesis Testing

Population Parameter(s):	Hypothesis Testing – Test Statistic	Sample Size
Mean (μ) (when n > 30 and n < .05N)	$Z_{st} = \dfrac{\left(\overline{X} - \mu_0\right)}{\left(\dfrac{\sigma}{\sqrt{n}}\right)}$	$\dfrac{\left[\left(Z_\alpha + Z_\beta\right)^2 \sigma^2\right]}{\Delta^2}$
Mean (μ) (when n < 30)	$t_{st} = \dfrac{\left(\overline{X} - \mu_0\right)}{\left(\dfrac{s}{\sqrt{n}}\right)}$	$\dfrac{\left[\left(Z_\alpha + Z_\beta\right)^2 \sigma^2\right]}{\Delta^2}$
Proportion (ϕ) (when $n\phi \ge 5$ and $n(1-\phi) \ge 5$)	$Z_{st} = \dfrac{\left(p - \phi_0\right)}{\sqrt{\left[\dfrac{p(1 - p)}{n}\right]}}$	-

Variance (σ^2)	$\chi^2_{st} = \dfrac{\left[(n-1)s^2\right]}{\sigma_0^2}$	-
Two Means ($\mu_1 - \mu_2$) (regardless of sample size when σ^2 is known)	$Z_{st} = \dfrac{\left[(\overline{X}_1 - \overline{X}_2) - \mu_0\right]}{\left[\left(\dfrac{\sigma_1^2}{n_1}\right) + \left(\dfrac{\sigma_2^2}{n_2}\right)\right]}$	$\dfrac{\left[\left(Z_\alpha + Z_\beta\right)^2 \left(\sigma_1^2 + \sigma_2^2\right)^2\right]}{\Delta_1^2}$
Two Means ($\mu_1 - \mu_2$) (when $30 \leq n_1 < .05N_1$ and $30 \leq n_2 < .05N_2$)	$Z_{st} = \dfrac{\left[(\overline{X}_1 - \overline{X}_2) - \mu_0\right]}{\left[\left(\dfrac{s_1^2}{n_1}\right) + \left(\dfrac{s_2^2}{n_2}\right)\right]}$	$\dfrac{\left[\left(Z_\alpha + Z_\beta\right)^2 \left(\sigma_1^2 + \sigma_2^2\right)^2\right]}{\Delta_1^2}$
Two Means ($\mu_1 - \mu_2$) (when $n_1 < 30$ or $n_2 < 30$)	$t_{st} = \dfrac{\left[(\overline{X}_1 - \overline{X}_2) - \mu_0\right]}{\left[\left(\dfrac{s_1^2}{n_1}\right) + \left(\dfrac{s_2^2}{n_2}\right)\right]}$	$\dfrac{\left[\left(Z_\alpha + Z_\beta\right)^2 \left(\sigma_1^2 + \sigma_2^2\right)^2\right]}{\Delta_1^2}$
Two Proportions ($\phi_1 - \phi_2$) (when $30 \leq n_1 < .05N_1$ and $30 \leq n_2 < .05N_2$)	$Z_{st} = \dfrac{\left[(p_1 - p_2) - \phi_0\right]}{\left[\left(\dfrac{p_1(1-p_1)}{n_1}\right) + \left(\dfrac{p_2(1-p_2)}{n_2}\right)\right]}$	-

Quick-Reference Guide for Chapter 5
Regression and Correlation

Correlation Coefficient: r_{xy}	$$r_{xy} = \frac{\Sigma AB}{\sqrt{(\Sigma A^2)(\Sigma B^2)}} \, ,$$ where $A = x - \bar{x}$ and $B = y - \bar{y}$
Regression Coefficients: b_1 and b_0	$$b_1 = \frac{\Sigma AB}{\Sigma A^2} \, ,$$ where $A = x - \bar{x}$ and $B = y - \bar{y}$ and $$b_0 = \bar{y} - b_1(\bar{x}).$$
Estimated Equation of Regression Line: $\hat{y}$	$$\hat{y} = b_0 + b_1 x$$
Coefficient of Determination: R^2	$$R^2 = \frac{(\Sigma AB)^2}{(\Sigma A^2)(\Sigma B^2)}$$

Quick-Reference Guide for Chapter 6
Nonparametric Statistics

Wilcoxon Rank-Sum Or Mann-Whitney Test	$$\mu_{R_a} = \frac{n_a\left(n_a + n_b + 1\right)}{2}$$ $$\sigma_{R_a} = \sqrt{\frac{\left[\frac{(n_a n_b)}{(n_a + n_b + 1)}\right]}{12}}$$ $$Z = \frac{\left(R_a - \mu_{R_a}\right)}{\sigma_{R_a}}$$
Sign Test	$\mu_s = .5n$, where n = (# of "+") and (# of "-"), $\sigma_s = \sqrt{.25n}$, and for n $\geq$ 10; $Z = \frac{\left(S - \mu_s\right)}{\sigma_S}$
Wilcoxon Signed-Rank Test	$$\mu_T = 0$$ $$\sigma_T = \sqrt{\frac{[n(n-1)(2n-1)]}{6}},$$ and if n $\geq$ 10, $Z = \frac{\left(T - \mu_T\right)}{\sigma_T}$
Kruskal-Wallis Test	$$K = \left[\frac{12}{n(n-1)}\right]\left[\Sigma\left(\frac{w_i^2}{n_i}\right)\right] - 3(n+1)$$ χ^2 distribution with Degrees of Freedom (DF), DF = n_s-1
Spearman's Rank-Correlation Test	$$\rho = 1 - \left[\frac{6\left(\Sigma d_i^2\right)}{n(n^2 - 1)}\right]$$ with $d_i = x_i - y_i$, where $x_i (y_i)$ = ranking of data point i of variable X(Y); $-1 \leq \rho \leq 1$

References

History of Statistics
Stigler, S.M., *The History of Statistics*, Cambridge, Mass.: Harvard University Press, 1986

Nature of Statistical Data and Scaling
Hildebrad, D. K., Laig, J.D., and Rosenthal, H., *Analysis of Ordinal Data*. Beverly Hills, Calif.: Sage Publications, Inc., 1977.
Reynolds, H.T., *Analysis of Nominal Data*. Beverly Hills, Calif.: Sage Publications, Inc., 1977.

Statistics for Laypersons
Cuzzort, R.P. And Vrettos, J.S., *The Elementary Forms of Statistical Reason*. St. Martin's, 1996
Hooke, R., *How to Tell the Liars from the Statisticians*. New York: Marcel Dekker, Inc., 1983.
Runyon, R.P., *Winning with Statistics*. Reading, Mass.: Addison-Wesley Publishing Company, Inc., 1976.

Graphs, Charts, and Summary Measures
Hamburg, M., *Basic Statistics: A Modern Approach*. New York: Harcourt Brace, Jovanovich, Inc., 1974.
Schmid, C.F., *Statistical Graphics: Design Principles and Practices*. New York: John Wiley & Sons, Inc., 1983.
Tufte, E.R., *The Visual Display of Quantitative Information*. Chesire, Conn.: Graphics Press, 1985.
Freund, J.E., *Mathematical Statistics*, 5th ed. Englewood Cliffs, N.J..: Prentice-Hall, Inc., 1987.

Elementary Probability and Ethics
Scheaffer, R.L. And Mendenhall, W., *Introduction to Probability: Theory and Applications*. Boston: Duxbury Press, 1975.
Mosteller, F., *Fifty Challenging Problems in Probability with Solutions*. Reading, Mass.: Addison-Wesley Publishing Co, Inc., 1995.
Hastings, N.A.J., and Peacock, J.B., *Statistical Distributions*. London: Butterworth & Co. (Publishers) Ltd, 1975.
Strait, P. T., *A First Course in Probability and Statistics with Applications*. Harcourt-Brace-Jovanovich, 1983.

Hogg, R. V., and Tanis, E. A., *Probability and Statistical Inference*, 2nd ed. Macmillan, 1983.

Brunk, H.D., *An Introduction to Mathematical Statistics*, 3rd ed. Herox, 1975.

Ross, S., *A First Course in Probability*, 3rd ed. Macmillan, 1988.

"Bayesian" Statistics

Motulsky, H., *Intuitive Biostatistics*. Oxford, 1995.

Kennedy, P., *A Guide to Econometrics,* 3rd ed.. Cambridge, MA., 1994.

Sampling

Cochran, W.G., *Sampling Techniques*, 3rd ed. New York: John Wiley & Sons, Inc., 1977.

Interval Estimation and Testing of Hypothesis

Freund, J. E., and Walpole, R.E., *Mathematical Statistics*, 4th ed. Englewood Cliffs, N.J.: Prentice-Hall, Inc., 1987.

Freedman, D., Pisan, R., and Purves, R., *Statistics*. New York: W.W. Norton &Co, Inc., 1978.

Maxwell, E. A., *Introduction to Statistical Thinking*, Englewood Cliffs, N.J.: Prentice-Hall, Inc., 1983.

Moroney, M.J., *Facts from Figures*. London: Penguin Books, Ltd., 1956.

Regression and Correlation

Draper, N.R., and Smith, H., *Applied Regression Analysis*, 2nd ed. New York: John Wiley & Sons, Inc., 1981.

Gujarati, D.N., *Basic Econometrics*, 2nd ed. McGraw Hill, 1988.

Harris, R.J., *A Primer of Multivariate Statistics*. New York: Academic Press, Inc., 1971.

Wonnacott, T. H., and Wonnacott, R.J., *Regression: A Second Course in Statistics*. New York: John Wiley & Sons, Inc., 1981.

Ezekiel, M., and Fox, K.A., *Methods of Correlation and Regression Analysis*, 3rd ed. New York: John Wiley & Sons, Inc., 1959.

Nonparametric Tests

Conover, W.J., *Practical Nonparametric Statistics*. New York: John Wiley & Sons, Inc., 1971.

Daniel, W.W., *Applied Nonparametric Statistics*. Boston: Houghton Mifflin Co. 1978.

Siegel, S., *Nonparametric Statistics for the Behavioral Sciences*. New York: McGraw-Hill Book Co., 1956.

Biostatistics and Health and Medical Applications/Problems
National Center for Health Statistics. (1979), *Monthly vital statistics report, annual summary.*
Motulsky, H., *Intuitive Biostatistics*. Oxford, 1995.
Pagano, M. And Gauvreau, K., *Principles of Biostatistics*. Duxbury, 1993.
Report of the Second Task Force on Blood Pressure Control in Children. (1987). *Pediatrics*, 79(1), 1-25.
Rosner, B., *Fundamentals of Biostatistics*, 4th ed. Duxbury, 1995.
Torok-Storb, et al. (1985). Subsets of patients with aplastic anemia identified by flow microfluorometry. *New England Journal of Medicine*, 312(16), 1015-1022.

Business Statistics
Aczel, A.D., *Complete Business Statistics*. Irwin, 1989.
Kohler, H., *Statistics for Business and Economics*, 3rd ed. Harper-Collins, 1994.

Subject Index

A

Addition Law, 16, 18

Alternative Hypothesis, 131

Autocorrelation, 169, 170

Axiom, 12, 15

Axiomatic Probability, 6, 12-19

B

Ballentine-Venn Diagram, 1-3

Bar charts, 32

Bayesian Analysis, 24, 29

Bayesian Theorem, 23

Bayes' Rule, 23

Bimodal, 40

Box-and-Whiskers Plot, 34, 43

C

Categorical Data, 189

Census, 4

Central Limit Theorem, 84, 85

Chi-square Distribution, 108, 109, 268-269

Classical Probability, 6-10

Clustering, 5

Coefficient of Determination, 168

Coefficient of Variation, 39

Conditional Probability, 19-24

Confidence Interval, 99, 105, 108, 110, 113, 173

Confidence Level, 105, 108

Continuous Probability Distribution, 89

Continuous Random Variable, 32

Convenience Sampling, 5

Correlation, 45, 54, 161, 204

Covariance, 45, 48, 51

D

Data (types), 2

- *Categorical,* 189

- *Nominal,* 189

- *Ordinal,* 189

Data Collection, 4

Degrees of freedom, 35, 97

Dependent Events, 14

Descriptive Statistics, 1, 31

Diagnostic/Lab Tests, 150

Durbin-Watson Statistic, 170, 274

Discrete Random Variable, 32

Distribution-free Testing, 189

E

Element, 12

Empirical Probability, 6, 10 – 11

Epidemiological Tests, 24

Error of Acceptance, 139

Event, 12

- *Dependent,* 14

- *Independent,* 14, 18

Expected Value, 45

Experiment, 4, 12

F

F Distribution, 143, 270-273

False Negatives, 26

False Positives, 26

Frequency Distribution, 31

G

Generalized Multiplication Law of
   Probability, 20

Grouped Data, 33, 36, 37

H

Heteroscedasticity, 170

Histograms, 32, 34, 81, 83, 88

Hypothesis Testing Procedure, 131-133,
   150, 172

I

Independent Events, 14, 18

Independent Samples, 191, 202

Inductive Statistics, 1

Inferential Statistics, 1

Interval of Prediction, 174

K

Kruskal-Wallis Test, 190, 202-203

Kurtosis, 34, 41

L

Left-tail Test, 131, 138, 149

Leptokurtic, 42

Lilliefors Test, 190

Linear Population, 165

M

Mann-Whitney Test, 190, 191

Matched-pairs, 194, 198

Maximum Error of the Estimate, 114

Mean, 34, 35

Median, 39

Mesokurtic, 42

Mode, 34, 40

Multimodal, 40

Multiplication Law of Probability, 15

Multiplication Principle, 7

N

Negative Posterior Probability, 27

Negative Predictive Value, 26

Negative Prior Probability, 25

Nominal Data, 189

Nonparametric, 189

Normal Distribution, 89

Null Hypothesis, 131

O

Observed Significant Level, 133

Ordinal Data, 189

P

P-value, 133

Parameter, 3

Parametric, 189

Park Test, 170

Percentile, 34, 39

Permutations, 8

Pilot Sample, 114

Platykurtic, 42

Population, 3, 165

Population Correlation, 55

Population Covariance, 51

Population Parameters, 34

Population Size, 34

Positive Posterior Probability, 27

Positive Predictive Value, 25

Positive Prior Probability, 25

Posterior Probability, 23

Post-Sampling, 102

Power, 139

Pre-Sampling, 100

Prevalence, 25

Prior Probability, 23

Probability

   - *Classical,* 6

   - *Empirical,* 6, 10-11

   - *Subjective,* 6, 11-12

   - *Axiomatic,* 6, 12-19

   - *Conditional,* 6, 19-24

Probability Distribution, 31, 89

Probability Theory, 5

Proposition, 12

R

Random Error "u", 169

Random Sampling, 4

Random Variable, 32, 45

Raw Data, 36, 38

Regression, 165

Regression Coefficients, 166, 173

Rejection Region, 131

Right-tail Test, 132, 138, 149

Robustness, 189

S

Sample, 2, 165

   - Random, 2

   - Representation, 2

Sample Correlation, 55

Sample Size, 35, 114-118, 148

Sample Space, 12, 33

Sampling, 4

   - Clustering, 5

   - Convenience, 5

   - Random, 2

   - Stratified, 4

   - Systematic, 5

Sampling Distribution, 81

   - of the Mean, 51, 55

   - of the Sample Proportion, 84, 85

   - of the Variance, 51

Sampling with Replacement, 8

Sampling without Replacement, 9

Scatter Plot, 52, 53, 56, 164, 168, 171, 175

Screening Tests, 24

Sensitivity, 25, 113

Set, 2

Sign Test, 190, 194-196

Significance Level, 131, 139

Simulation, 4

Skewness, 34, 40

Spearman's Rank-Correlation Test, 190, 204-205

Specificity, 25

Standard Deviation, 34, 38

Standard Error of the Mean, 83

Standard Normal Distribution, 90, 265-266

Statistic, 3

Statistical Hypothesis, 131

Statistical Significance, 172

Stem-and-Leaf Display, 34, 44

Stratified Sampling, 5

Strong Law of Large Numbers, 11

Subjective Probability, 6, 11-12

Sub-set, 2, 3

Systematic Sampling, 5

T

t-Distribution, 97, 98, 267

Theorem, 12, 15-19, 20-24

Tolerable Error, 114

Total Probability Rule, 22

Type I Error, 138

Type II Error, 138

Two-tail Test, 132, 139, 149

U

Ungrouped Data, 33, 36, 37

V

Variance, 34, 37, 45, 46

Venn Diagram, 13, 14, 16, 22, 25

W

Wilcoxon Signed-Rank Test, 190, 198-
199

Wilcoxon Rank-Sum Test, 190, 191

X

χ^2-Distribution, 108, 109, 268-269

Z

Z-Distribution, 90, 91, 265-266